CPS SERIES IN
PHILOSOPHY OF SCIENCE

Center for Philosophy of Science
University of Pittsburgh

CPS PUBLICATIONS IN PHILOSOPHY OF SCIENCE

Center for Philosophy of Science
University of Pittsburgh

EDITED BY

Adolf Grunbaum
Larry Laudan
Nicholas Rescher
Wesley Salmon

THE LIMITS
OF LAWFULNESS

Studies on the Scope and Nature of
Scientific Knowledge

EDITED BY

Nicholas Rescher

University Press of America

Copyright © 1983 by the
Center for Philosophy of Science
University of Pittsburgh

University Press of America,™ Inc.

4720 Boston Way
Lanham, MD 20706

3 Henrietta Street
London WC2E 8LU England

Library of Congress Cataloging in Publication Data

Main entry under title:

The Limits of lawfulness.

(CPS publications in philosophy of science)
Includes bibliographical references and index.
1. Science—Philosophy. I. Rescher, Nicholas.
II. Series.
Q175. L54 ǂ 1983   501   83-6872
ISBN 0-8191-3176-8
ISBN 0-8191-3177-6 (pbk.)

CPS PUBLICATIONS IN PHILOSOPHY OF SCIENCE
Co-published by arrangement with the
Center for Philosophy of Science
University of Pittsburgh

CONTENTS

PREFACE

These essays originated from an interdisciplinary conference on "Limits of Science" held in Pittsburgh on October 15-16, 1982, under the sponsorship of the University of Pittsburgh's Center for Philosophy of Science. This conference, which was supported by the Richard King Mellon Foundation, brought together philosophers, historians, and scientists. It is the hope of the sponsors of the conference, and of the editors of this volume which is its outgrowth, that these discussions will serve both to inform and to stimulate further reflections on this important topic. The theme is, of course, so vast that the discussions presented here can do little more than scratch the surface.

A.
HISTORICAL PERSPECTIVE

NEGATIVISM SESQUICENTENNIAL

Stephen G. Brush

O ne hundred years ago Ernst Mach proposed his notorious principle of the economy of thought in scientific research. Mach urged scientists to abandon the goal of discovering the intrinsic nature of bodies apart from our perceptions of them, and of determining causal relations between phenomena apart from the mathematical description of their correlations. Atomic models may be temporarily useful in representing the results of observations but it is futile, he argued, to attribute reality to them (or indeed to any substance apart from its perceived attributes); and to the question, "how is it possible to explain feeling by the motions of the atoms of the brain?" Mach replied: *certainly this will never be done* (1943: 208).

Mach's insistence on limiting the possible range of reliable knowledge to the domain of human sense perception gave a powerful impetus to a movement started 50 years earlier by August Comte. Known, quite inappropriately, as *positivism*, this movement glorified science yet attempted to limit its scope so severely as to discourage some scientists from seeking knowledge that was in fact attainable. Of the two aspects of positivism—the claim that the methods of the physical sciences should be applied to all inquiry, sometimes called "scientism," and the claim that scientific knowledge is limited to direct sense perception—I shall focus on the latter. Despite the apparent demise of positivism in the last few decades, we continue to be assaulted by assertions that "science can never find out X" or "scientists can never do Y." Negativism persists despite an astonishing sequence of successful discoveries of supposedly unknowable X's and performances of supposedly impossible Y's. Recent examples are a molecular theory of emotion (forbidden by Mach himself, as mentioned above) and several births of "test-tube babies."

After examining Comte's own statements about the limits of scientific knowledge

3

I will discuss four historical examples of attempts to impose such limits, beginning with Comte's proscription of extrasolar-system astronomy and ending with current experiments inspired by the Einstein-Podolsky-Rosen paradox. I have chosen these examples, for the sake of symmetry, to illustrate claims that we cannot know about events at (a) very large distances; (b) very small distances; (c) very long times (in the past); (d) very short times.[1]

I

In his scheme of three stages — theological, metaphysical, and positive — Comte describes the desired third stage as one in which

> the mind has given over the vain search after Absolute notions, the origin and destination of the universe, and the causes of phenomena, and applies itself to the study of their laws — that is, their invariable relations of succession and resemblance. (1855: 26; original French version given in Appendix A)

As examples of successful science he points to Newton's theory of gravity and Fourier's theory of heat. In the former, we must abstain from asking about the *cause* of gravity and simply accept it as a fact:

> As to what weight and attraction are, we have nothing to do with that, for it is not a matter of knowledge at all. (1855: 29; Appendix B)

I presume Comte would thus reject not only mechanistic explanations such as Le Sage's hypothesis of ultramondaine corpuscles, but anything like Einstein's general theory of relativity.

As for heat, Comte praises Fourier for giving us the mathematical laws governing the phenomena, without inquiring into the *nature* of heat, which is unknowable. Thus the entire development of the kinetic-molecular theory of heat, which was one of the major achievements of 19th century physics (Brush, 1976), was an illegitimate excursion beyond Comte's posted limits of scientific knowledge. (Comte did not, however, reject the atomic structure of matter.)

Is it unfair to criticize Comte's philosophy on the basis of the later development of science? Not at all. Comte himself wrote that the Positive Method can be judged only in action — "it can not be looked at by itself, apart from the work on which it is employed." (1855; 34; App. C) Thus if the Positive Method proscribes a topic of inquiry which can in fact be successfully pursued, this must count as a serious objection to the Method.

A recurring theme in the early chapters of Comte's *Course of Positive Philosophy* is the impossibility of knowing anything about the universe beyond our own solar system. We don't even know the distances of any stars. Our measurements of their positions are necessarily inaccurate because their light is refracted by the earth's atmosphere before it reaches us; we cannot make a precise correction for this refraction because we "do not and can not know" the variation of density in the atmosphere. (1855: 142; App. D) We cannot assume that the law of gravity applies to the stars because

> there can be no positive science apart from phenomena, and of the phenomena of the universe beyond our own system we are not in scientific possession. (1855: 168; App. E)

As J. P. Nichol, a British astronomer, pointed out in his notes to Harriet Martineau's English version of the *Positive Philosophy*, this statement had already been refuted by the analysis of the motions of double stars, which conform to Kepler's laws. (1855: 168)

The physical nature of heavenly bodies is even more inaccessible than the dynamics of their motion; Comte flatly asserts that we will never be able to study the temperature or chemical composition of the stars. (App. F) So, he concludes,

> the study of the universe forms no part of natural philosophy. (1855: 185; App. G)

Likewise, the limits of cosmogony are set at the stage when the fluid solar nebula was rotating on its axis; Comte is willing to discuss what happened after that stage (and even fancied that he could provide a new confirmation of Laplace's nebular hypothesis) but admonished his readers not to speculate about earlier events and chastised Descartes for doing so.

Comte was equally positive that no answers could ever be obtained to other questions that intrigued 19th-century scientists. Physicists should stop speculating about "agents" of the phenomena of heat, light, electricity, and magnetism, i.e., ethers and imponderable fluids—not because they are non-existent, but because their existence can neither be proved nor disproved. He doubted that it made any sense to talk about heat apart from a warm body, or light apart from a luminous body; in any case, it would never be possible to reduce the phenomena of light to any other (Comte is perhaps reflecting the pessimistic views of many late 18th century physicists; see Heilbron 1982: 63–65).

Moreover, Comte warned, no one should even try to explain the colors of bodies—

> The so-called explanations, about the supposed faculty of reflecting or transmitting such and such a kind of rays, or of exciting such and such an order of ethereal vibrations, in virtue of certain supposed derangements of the molecules, are more difficult to conceive than the fact itself ... Nobody now tries to explain the specific gravity proper to any substance or structure: and why should we attempt it with regard to specific color, which is quite as primitive an attribute ... The field of inquiry is vast enough, without any such illusory research as this. (1855: 235; App. H)

Finally, Comte excludes theoretical chemistry from science:

Every attempt to refer chemical questions to mathematical doctrines must be considered, now and always, profoundly irrational, as being contrary to the nature of the phenomena. (1855: 256; App. I)

James Clerk Maxwell (1876), commenting on the work of J. Willard Gibbs which successfully transgressed this limit, surmised:

The warning which Comte addressed to his disciples, not to apply dynamical or physical ideas to chemical phenomena, may be taken, like several other warnings of his, as an indication of the direction in which science was threatening to advance.

Anyone familiar with the history of modern physical science knows that all of Comte's negative claims have been falsified — not in the sense that we have established the *true* answers to his forbidden questions, but rather than we have accepted the questions as legitimate ones for science to consider, and have found reasonably satisfactory working hypotheses about the nature of gravity, heat, light, the molecular reasons for the color of bodies, and the mathematical laws of chemical structure and reactions.

With such a disastrous record of failure, one might wonder why Comte's positivism deserves any serious consideration by historians and philosophers of science. For the historian, negative influences on scientific progress call for investigation; in fact, a recurrent question is: what caused the decline of French theoretical physics in the 19th century? Or, if we wish to avoid the pitfalls of the "Whig interpretation of the history of science," we should ask why French scientists rejected certain research problems and methods, without assuming that they were foolish to do so merely because those problems and methods now seem to be in the mainstream of modern physical science. Can we blame the positivists for failing to anticipate the discoveries of Maxwell and Einstein?

For the philosopher of science Comte's strictures may still have some interest when abstracted from their 19th century context. He insisted that successful prediction is the goal of science:

all science has prevision for its end: an axiom which separates science from erudition, which relates the events of the past, without any regard to the future. (1855: 135, App. J).

From this viewpoint, cosmology is not a science — it deals with phenomena that happened so long ago and far away that there is no way to bring them into the laboratory for controlled experimentation; and it is futile to predict the future since we will not live long enough to check the results. Similarly, as Karl Popper used to argue, evolutionary biologists can only try to explain singular events, they cannot establish laws or make testable predictions. Precisely this argument is now being used by religious fundamentalists to justify injecting creationism into public school science classes: any statements about the origin of the human race, or of the earth, go beyond the limits of scientific knowledge and therefore the views of modern biologists, geologists and astronomers should not be allowed to displace those of theologians.

Laudan (1971), in his analysis of Comte's methodology, points out that Comte did not always restrict predictions to future events; they could include events that have already happened but are previously unknown to us, such as eclipses in antiquity. Laudan credits Comte not only with giving serious attention to "retrodiction" but, more generally, with establishing the idea that testability rather than certainty should be the proper criterion of demarcation between scientific and nonscientific statements. Laudan also makes the interesting claim that Comte avoided some of the pitfalls into which Popper and the logical positivists tumbled because of their rigid definitions of meaning and verifiability.

Laudan also suggests that Comte's attitude toward hypotheses is somewhat more liberal than the above-quoted statements suggest; in particular, in his *System of Positive Polity* (1851) Comte "explicitly endorses the atomic theory" (Laudan, 1971: 49). But the passage quoted by Laudan is preceded and followed by warnings against taking atoms too seriously:

The ultimate structure of bodies must always transcend our knowledge. But while studying their properties

and now comes the passage quoted by Laudan

it is consistent with sound reasoning to make use of any hypotheses that will assist thought, provided always that they be not inconsistent with what we know of the phenomena.

Comte continues with what might well be called an "endorsement":

Now the Molecular hypothesis satisfies both these conditions in all inorganic researches, and especially in Physics, where it is favoured by the increased prominence of induction and experiment. While studying the general properties of matter, we find it useful to suppose them inherent in the smallest particles which the mind can conceive. Taking this as our starting-point, it becomes easier to realise the essential permanence of these various properties, the alterations in them being only those of degree. But while this relative mode of regarding the atomistic hypothesis is the philosophical justification for its employment, it is inconsistent with the belief in its absolute reality; and indeed it suggests limits to which it should be restricted. It is mere blind imitation to introduce it into Biology when the first principles of the Science are essentially synthetic, and therefore wholly alien to it. Even in Chemistry its strict application is limited; the properties there considered being too complicated and too variable to be attributed with any good result to unchangeable atoms. (p. 421)

So, Comte says in a passage Laudan does quote, the value of atomic theory is impaired by "our tendency to endow all subjective creations with objective existence, as though they represented some external reality." (Ibid.)

Comte thus agrees with Mach's view that the atom is no more than a useful tool for describing phenomena, and has no intrinsic reality apart from our own thoughts and sensations. A similar view, applied not to atoms but to subatomic particles, has been revived in modern physics. We can indeed "see" an atom but only by probing it with electrons or photons. We have nothing suitable to enable us to see electrons and photons; we can determine their existence in a definite state but only if we do not try to pin them to a precise location in space or time. Quantum mechanics imposes a limit on our knowledge of the instantaneous position of a particle, and leads us to wonder whether the particle really *has* an instantaneous position apart from our measurement of it. Perhaps the limit of knowledge coincides with the limit of reality.

II

Comte knew that astronomers had been trying to determine stellar distances for two centuries with no reliable results; but less than a decade after he published his declaration about the futility of studying the stars, the first successful measurements of stellar parallax were made, by F. W. Bessel and others. Bessel announced in 1838 that the star 61 Cygni changes its apparent position as seen from different points along the earth's orbit, indicating that its distance is about 10^{14} kilometers (60 million million miles, or 10 light years). This is within 10% of the modern value, and there seems no reason to doubt that this kind of measurement, when it can be made, gives us reliable knowledge about stellar distances.

Did this advance in stellar astronomy cause Comte to change his views about the limits of astronomical knowledge? On the contrary, he betrayed his contempt for science by asserting, in *Positive Polity*, that astronomers should limit themselves to studying primarily the earth, and other bodies in the solar system only insofar as their behavior affects the earth. This astonishing statement is consistent with the general doctrine that one should study each science only to the extent needed to understand the next one "above" it in a hierarchy topped by sociology. Contrary to Laudan's assertion that "Comte values theories and laws for their own sake, not because they make possible a world of better gadgets" (1971: 36), Comte argues that all of science should be pared down to the parts that are relevant to humans.

> When all the heavenly bodies were supposed to be connected with the earth, or rather subordinate to it, it was reasonable that none should be neglected. But now that the Earth's motion is known to us, it is not necessary to study the fixed stars, except so far as they are required for purposes of terrestrial observation, and astronomy properly so called may be reduced to the study of the solar system. Even supposing it possible to extend our investigations to other systems, it would be undesirable to do so. We know now that such investigations can lead to no useful result: they cannot affect our views of terrestrial phenomena, which alone are worthy of human attention. (1851: 312-13).

Even the extension of the solar system by the discovery of Neptune in 1846 was rejected: Comte ridiculed the "insane enthusiasm" of astronomers and the public at a so-called discovery, which, even supposing it genuine, could have no real interest except for the inhabitants of Uranus. (ibid.)

III

The first determination of the chemical composition of stars, an impossible feat according to Comte, took a little longer. In 1859 Gustav Kirchhoff and Robert Bunsen showed that chemical elements could be identified by characteristic spectral lines (light absorbed or emitted at definite wavelengths). William Huggins and W. A. Miller used this method to identify elements in stars in the early 1860s. Huggins also showed, in 1868, that small displacements in spectral lines could be explained by the Doppler effect, and estimated that the star Sirius is moving away from us at more than 20 miles per second.

The analysis of stellar spectra provided the chief basis for the spectacular growth of a new science, astrophysics, in the late 19th and early 20th centuries. It was possible to estimate not only the chemical composition of the surface of a star but also the temperature and pressure in its outer layers, by applying the quantum statistical theory of ionization and excitation of atoms. During the same period, the direct parallax method for estimating stellar distances was extended and supplemented by indirect techniques, notably the use of the period-luminosity correlation for Cepheid variables discovered by Henrietta Leavitt in 1908. With this method astronomers could assign distances on the order of hundreds of thousands or millions of light years to galaxies outside our own. The Doppler shifts in spectral lines could be used to estimate the speeds at which these galaxies are approaching or receding from us.

Astrophysics and galactic astronomy are extremely anti-positivist sciences, not only because they deal with topics proscribed by Comte himself, but because they always seem to be violating some reasonable limit on what we can know about the universe. For example, cosmological theories based on E. P. Hubble's discovery of a correlation between the distances and recession velocities of galaxies deal with enormous extrapolations of the "expanding universe." Between 1926 and 1952 it appeared that the extrapolation backwards in time, based on accepted values for distances and velocities, would lead to a singular state (all matter concentrated at a point) only a couple of billion years ago, contradicting radiometric evidence that the earth itself is older than that. The difficulty was avoided by a correction of the distance scale which doubled the "age" of the universe, but allowed skeptics to question the reliability of *any* estimate of the distances of galaxies. Similarly, the use of spectral line shifts to estimate distances gave such implausible results for "quasars" that many astronomers began to wonder whether those shifts might be due to something other than the Doppler effect. Currently fashionable interpretations of the "big bang" and of quasars attribute both to enormous black holes which can be viewed as ultimate limits of knowledge—the theory postulates a "cosmic censorship" or "retrodiction barrier" which prevents us from getting any information about certain regions of space and time.

Astrophysicists and cosmologists take dangerous epistemological risks—one might argue that they have already transgressed the limits of scientific knowledge and have been operating in a land of metaphysical fantasy for nearly a century. In any case it is interesting to note that the French and the Germans, who were world leaders in astronomy in the early 19th century, seem to have abandoned this field to the British and Americans (Brush, 1979: 45-47). Is this because the French and Germans took too seriously the alleged limits on scientific knowledge proclaimed by Comte and Mach?

IV

John Herivel pointed out several years ago that French scientists made no significant contributions to "the three major creative achievements in theoretical physics in the second half of the nineteenth century: the advances in electricity and magnetism culminating in the theory of Maxwell, and the creation of thermodynamics and the kinetic theory of gases"—despite the fact that decisive contributions in those and closely related areas had come from French scientists in the early part of the century (Sadi Carnot, Coloumb, Poisson, Ampère, Laplace). Herivel attributed this decline in French theoretical physics to the dominance of the positivist tradition, articulated by Comte and exemplified by Fourier. Rejecting the "mechanical" tradition associated by Laplace, Ampère, and Poisson, which emphasized the reduction of phenomena to molecular forces and motions, Fourier asserted that the phenomena of heat could not be reduced to mechanics, and Comte argued more generally, as we have seen, that one type of phenomenon could not be reduced to another or to ultimate causes; science should be limited to the search for empirical laws, and the atomic hypothesis is only a "logical artifice" which, though useful, does not touch reality. (See Laudan, 1971: 46-52 for further discussion). Moreover, Comte "categorically denied the possibility of explaining physical phenomena in terms of chance or probability" (Herivel, 1966: 125) and denounced the "Irrational approval given to the so-called calculus of chances" because it presumed the complete absence of law (Comte, 1851: 381).

According to Herivel, the decisive advances in thermodynamics, kinetic theory, and electromagnetism after 1840 were achieved by the use of strategies exactly contrary to those advocated by Comte and Fourier: the development of specific hypotheses about the nature of heat, elaboration of hypothetical and even fantastic mechanical models, and exploitation of statistical techniques. In France, research was dominated by "strong emphasis on observation and scientific law as opposed to theory or speculation"—an approach that, in the case of the best workers such as Regnault, led to solid experimental results but no theoretical understanding. (Herivel, 1966: 129) Herivel's thesis has encountered some skepticism (especially among those anxious to defend the reputation of French science) but it has not been refuted; I think Herivel has made a *prima facie* case for the charge that attempts to limit scientific knowledge, if taken seriously, are likely to damage the progress of science.

Although Clausius, Maxwell and Boltzman were generally optimistic about the prospect of discovering molecular properties, their use of statistical methods was itself justified by an assertion about the practical limits of knowledge about molecular motions. Like Laplace who followed his famous statement on determinism by saying that we cannot obtain complete knowledge and even if we could, we cannot do the immensely involved calculations needed to predict the future state of the system, Maxwell and his disciples often stated that we cannot deal individually with the huge number of molecules in a cubic centimeter of gas. But even that limit now seems to be unduly pessimistic; with modern computers it was possible as early as 1956 to compute the motions of a few hundred molecules accurately enough (assuming quantum effects are negligible) to account for phenomena such as solid-fluid transitions (Alder and Wainwright, 1957). This would seem to refute the statement, still frequently heard, that phenomena on one level cannot be reduced to those on a lower level without introducing new laws.

The great success of the mechanistic-atomistic program in the period 1850–70 is reflected in a remarkable pronouncement on the limits of scientific knowledge by Emil du Bois Reymond in 1871. Du Bois Reymond was one of the first scientists to stress the significance of Laplace's assertion that an intelligence which could comprehend at one instant all the positions, velocities, and force of the atoms of the universe, and had unlimited computing power, could reconstruct the entire past and future. (Cf. Brush, 1976: 584). Du Bois Reymond argued that while the human mind is unlikely to approach this degree of perfect knowledge, we are dealing here only with a matter of degree, not a rigid limitation; perhaps a mind like that of Newton, being in possession of the necessary basic equations, differs less from Laplace's superintelligence than "the mind of an Australian or of a Fuegian savage differs from the mind of Newton." (p. 20). So we should consider that everything in principle accessible to Laplacean determinism is *within* the limits of scientific knowledge; Du Bois Reymond calls this "astronomical knowledge."

For Du Bois Reymond, there is no impassable barrier to knowledge at the boundary between inorganic and organic matter; he sees the former as a system of molecules in stable equilibrium, the latter as a system in dynamic equilibrium maintained by the influx and transformation of energy. They differ just as a building differs from a factory. We can possess astronomical knowledge of plants, animals, and humans. But we cannot understand *consciousness*, even when we can trace the motions of all the atoms in the brain. This is one absolute limit on our knowledge; the other is that we cannot understand the true nature of matter and force. Having claimed for mechanistic science a relatively wide domain, Du Bois Reymond ends his address thus:

> as regards the enigma what matter and force are, and how they are to be conceived, (the scientist) must resign himself once for all to the confession—IGNORABIMUS!

Du Bois Reymond's viewpoint suggests another side of the term "positivism" often found in popular writings on science: the claim that there is no valid knowledge other than scientific knowledge, and that the methods of the natural science should be extended to all other domains of inquiry. The proposition that one can explain, predict, and control such phenomena as the evolution of species and human behavior, using a purely mechanistic approach, is repulsive to many people. When they proclaim limits of scientific knowledge, they may characterize their standpoint as a *rejection* of positivism. This is the "computers can never think" doctrine; rather than being based on confidence that the limits will never be violated, it seems to be inspired by a fear that they will be. Thus the bold statements of scientists like T. H. Huxley and John Tyndall in the 1870s, that biological and psychological phenomena would soon be reduced to mechanical principles, triggered widespread fears of the success of "materialism." Worst of all, the scientists dared to subject divine power itself to experimental investigation in the tests of the effectiveness of prayer proposed by Tyndall and Francis Galton (Brush, 1978: 77-90).

V

As early as 1870, quantitative estimates of the size, mass, and average speeds of atoms were available, and the numbers were consistent with theoretical interpretations of several different observable properties of matter. (Brush, 1976: 75-78). Yet in the 1890s, Mach and Ostwald were able to persuade many scientists that the atomic theory should be abandoned, or at best relegated to a heuristic description. It is sometimes argued that the rejection of atomism was not the result of philosophical prejudice but was objectively justified by the situation in physical science at that time—that, in the terminology of Imre Lakatos, the kinetic-molecular research programe of Boltzmann and his colleagues was "degenerating." (Clark, 1976; for a refutation of this view see Nyhof, 1981) According to this view, one might say that positivism helped to disencumber science of superfluous hypotheses and thus clear the way for theories more consistent with experimental results. Similarly, Mach's 20th-century followers in the Vienna Circle pointed to his critique of Newtonian absolute space and time as clearing the way for Einstein's relativity, which is thus seen as a positivistic theory based only on observable quantities. (Einstein's friend and biographer Philipp Frank, one of those responsible for promoting this view of Einstein, was at least honest enough to record his own surprise when he learned, in 1929, that Einstein was opposed to positivism; see Frank, 1941: 30-57 and 1947: 215).

While there may be some basis for this view—the proliferation of mechanistic models of ether and atoms in the 19th century did not seem to advance science very much—it can hardly be maintained that the pioneers of 20th century physics were positivists. Both Einstein and Planck repudiated Mach's philosophy and advocated a realist conception of nature (Holton, 1968; Toulmin, 1970).

A striking example of the damage that can be done by positivist restraints on scientific theorizing is the failure of Pierre Curie and André Debierne to recognize radioactive transmutation of elements, recently discussed by Marjorie Malley

(1979). In 1901, with the evidence from their own experiments,

> they could have easily concluded that the radium emanation was a gas evolved by radium, which clearly was the hypothesis they were testing.... Yet they rejected Rutherford's conclusion that thorium (and by implication radium) evolved a radioactive gas, because they thought he had strayed beyond the established facts by endowing his working hypothesis with that status of reality: "as one can easily conceive other satisfactory explanations it seems premature to us to adopt any theory whatsoever." Their refusal to commit themselves to a specific hypothesis (because of Curie's positivism) or to attach much importance to the question of the nature of the emanation and excited activity . . . stalemated their research." (Malley, 1979: 217).

Despite having a head start in this area of research, the French scientists fell behind Rutherford because of their reluctance to make bold hypotheses. Not only did they miss making the discovery of nuclear transmutation, one of the major breakthroughs of 20th century science, they were quite slow to accept it when Rutherford and Soddy announced their conclusions a year later.

Malley notes that Marie Curie's earlier papers, written alone, were more speculative but her collaboration with Pierre led her to adopt her husband's positivism. Thus Pierre's negative influence persisted beyond his death in 1906; Marie's research group "eschewed basic theoretical questions. Busy work became the main pursuit in Paris until after World War I." (Malley, 1979: 223)

In spite of the deadening influence of positivism, scientists did of course eventually accept the existence of atoms as well as radioactive transmutation and it is worth noting that another Frenchman, Jean Perrin, was largely responsible for this. Perrin used Einstein's theory of Brownian movement to obtain an independent confirmation of the earlier estimates of atomic diameters, and showed that some of the results predicted by Einstein could be considered a direct indication of the graininess of matter. Thus Perrin established what had been a matter of speculation for two millenia: the atomic nature of matter. Curiously enough he rarely gets credit for this remarkable achievement: few of the books that discuss atoms mention the man who proved that they exist!

Since the most important results on Brownian movement were announced in 1908 (Brush, 1976: 693-701; Nye, 1972), I propose that we celebrate in 1983 the 75th anniversary of this anti-positivist milestone, the proof of the existence of atoms.

VI

The most vociferous limiters of scientific knowledge today are the religious fundamentalists who want to force "creationism" into public school science classes. Rejecting the consensus of scientists, that present forms of life have evolved from simpler ones over billions of years, creationists assert that the entire universe was created in 6 days about 6000 years ago, and that evolution has never changed one "kind" of organism into another. As was shown in the recent trial in Arkansas, their use of the word "kind" is a clear indication of the dependence of creationism on a particular religious creed. Lacking any scientific evidence for their theory, the creationists argue that the whole question of origins is beyond the limits of scientific knowledge; since we have no way to observe what happened in the distant past, one theory is as good as another. Positing that evolution and creationism are the only two possible alternatives, they urge that their doctrine deserves equal time with evolution.

The creationists bolster their position with the arguments of Karl Popper, who called Darwinism a metaphysical research programme, not a testable scientific theory. It can only try to explain what has already happened, but cannot make predictions about the future. (Popper, 1976: 167–80) According to Popper's famous criterion, a theory which can never be falsified does not deserve to be called "scientific"—falsifiability provides the line of demarcation between science and pseudoscience.

By defining "prediction" so narrowly as to exclude Darwinian evolution, Popper and the creationists must also put geology and astronomy in the category of pseudoscience. Since the scientific community does accept geology and astronomy as legitimate sciences, something must be wrong with Popper's falsifiability criterion. Popper himself has realized this and retracted his statement that Darwinism is not a scientific theory. (Popper, 1978, 1980) In fact evolutionary theories do make predictions, in the sense that they specify the result of observations that have not previously been made, pertaining to populations rather than individual organisms. (Williams, 1982) The failure to observe predicted transitional organisms in the fossil record, frequently used by creationists as an argument against evolution, shows that the Darwinian gradualist version is indeed falsifiable and has led many biologists to adopt a different version, "punctuated equilibrium." The evidence that evolution has taken place is overwhelming, even though the mechanism of evolutionary change is still in doubt.

Nevertheless there is still some plausibility in the claim that we can never know what happened in the distant past, when events may have occured unlike those we can now observe. I would therefore like to present two examples of cosmogonic hypotheses that led to definite predictions—one confirmed, the other falsified.

The first is the "big bang" theory of the origin of the universe, developed by George Gamow and his colleagues. According to this theory the extremely high temperature radiation of the primeval "fireball" would have cooled off while spreading out to fill the universe more or less uniformly. The present temperature of this "black body" background radiation must be about $5°K$, according to an estimate published in 1949 by Alpher and Herman. This "prediction" was dramatically confirmed in 1965 by Penzias and Wilson, and provides one of the strongest pieces of evidence for the big bang theory (although there are still some difficulties with that theory). The award of the Nobel Prize to Penzias and Wilson is evidence that the scientific community does not regard theories of the origin of the universe to be beyond the limits of scientific knowledge.

My other example (Brush, 1982) comes from selenogony: Harold Urey's theory that the moon was formed elsewhere in the solar system and later captured by the earth. Urey argued that the moon has been frozen since its capture, and records on its surface valuable information about the past history of the solar system—information that has been destroyed on the earth's surface by geological processes. One alternative had been proposed by G. H. Darwin (son of Charles Darwin): the moon was born by fission from the earth. Urey pointed out that if the moon were merely a piece of the earth it would not be scientifically worthwhile to explore it extensively.

Urey made several specific predictions from his capture theory, for example that significant amounts of water would be found on or near the lunar surface. Proponents of the fission theory, in particular John O'Keefe, also made predictions, for example that the abundance of nickel and similar elements would be much less than the solar system average.

The test of Urey's hypothesis (the Apollo program) cost $20 billion, though most of that amount could be charged to the political objective of re-establishing technological superiority over the USSR. The result was that the hypothesis was falsified, while O'Keefe's was at least partly confirmed. Urey himself abandoned his theory and leaned toward the Darwinian fission theory. (O'Keefe & Urey, 1977) Other scientists say that none of the pre-Apollo theories of the moon's origin is still tenable, and several rather artificial new hypotheses have been proposed.

VII

Atoms exist, and we can even see them with the field emission microscope (Müller, 1956)—yet they are not atoms in the original sense of being indivisible. The limits of scientific knowledge have moved down in the scale of distances to the electron and photon, and the question is no longer simply whether we can find out what goes on in very small parts of space, but whether we can do so for very small intervals of time. Can we know the instantaneous position and velocity of an electron? According to the indeterminacy principle of Heisenberg, we can determine one or the other but not both at the same time.

Einstein, as everyone knows, objected to the idea that atomic behavior is fundamentally random. This is one possible interpretation of Heisenberg's principle. But Einstein disliked even more, I think, the other interpretation: that atomic behavior has no independent reality apart from our observations.

To illustrate his objection, Einstein posed the following question: suppose a ball is in either of two closed boxes with equal probability. Assuming there is only one ball in the system, then by looking in box #2 you can establish whether or not the ball is in #1. Now according to common sense realism, the ball really was or was not in #1 before you looked in #2—yet according to quantum mechanics, if the ball is an electron, you cannot say that it was *in reality* in #1 or #2 before you looked. (Fine, 1979)

The more complicated version of this thought experiment, presented in a paper published under the names of Einstein, Boris Podolsky, and Nathan Rosen (1935), features two particles that interact briefly and then separate to an indefinitely great distance. If the interaction is such that a certain quantity like total momentum is conserved, then measurement of a property of one particle immediately determines that property for the other. In the modern version, an atom in an excited state decays to the ground state by emitting two photons. Since the angular momentum of the atom is known to be zero in both initial and final states, the spins of the two photons must be in opposite directions. Einstein, Podolsky, and Rosen argue that the spin of each photon must really exist before it is measured, since it can be determined by measuring the spin of the other photon, and since the other photon may be very far away this measurement cannot possibly disturb the first one. Since quantum mechanics denies independent reality to the spin of each photon before the measurement, it must be incomplete.

To me it seems clear that Einstein's argument is valid—that electrons and photons have properties before we measure them. But it now appears that I am mistaken. Several experiments conducted in the past decade have shown that quantum mechanics is right and Einstein is wrong. We are faced with the unhappy choice of abandoning either realism (the existence of the world apart from our perceptions) or "locality" (the postulate that effects cannot be transmitted instantaneously from one place to another). The only remaining question is whether information can be transmitted from one photon to the other at the speed of light (to be tested in the Aspect experiment now in progress). If not, we would apparently have an experimental disproof of realism—which would at least refute the claim that the validity of the philosophical foundations of science lies beyond the limits of scientific knowledge (cf. Kyburg's remarks at this conference)

If quantum mechanics is the most complete description of reality that is possible (contrary to Einstein, Podolsky, and Rosen) then we would have a well-defined limit on a certain kind of scientific knowledge. But this limit has not been imposed *a priori* by a philosopher or a philosophically-minded scientist; it is forced on us by the progress of science itself, its precise nature is established by experiment, and it is subject to revision by future discoveries. That, I would argue, is the only kind of limit that science can accept.[2]

Notes

1. Note that I deal only with limits on what science *can* know, not what it *should* know.

2. This paper is based in part on research supported by the History and Philosophy of Science Program of the National Science Foundation. I thank Frederick Suppe for numerous suggestions, some of which I have used here.

Bibliography

Alder, B. J. and Wainwright, T. E. 1957. "Phase Transition for a Hard Sphere System," *Journal of Chemical Physics*, vol. 27, pp. 1208-9.

Alpher, R. A. and Herman, R. C. 1949. "Remarks on the Evolution of the Expanding

Universe," *Physical Review*, series 2, vol. 75, pp. 1089-95.

Brush, Stephen G. 1976. *The Kind of Motion We Call Heat: A History of the Kinetic Theory of Gases in the 19th Century* (Amsterdam: North-Holland).

Brush, Stephen G. 1978. *The Temperature of History: Phases of Science and Culture in the Nineteenth Century* (New York: Burt Franklin).

Brush, Stephen G. 1979. "Looking Up: The Rise of Astronomy in America," *American Studies*, vol. 20, pp. 41-67.

Brush, Stephen G. 1982. "Nickel for your Thoughts: Urey and the Origin of the Moon," *Science*, vol. 217, pp. 891-98.

Clark, Peter. 1976. "Atomism versus Thermodynamics," in *Method and Appraisal in the Physical Sciences*, ed. by Colin Howson, pp. 41-105 (New York: Cambridge University Press).

Comte, Auguste. 1830-38. *Cours de Philosophie Positive*, Volumes I, II, III (Paris: Bachelier).

Comte, Auguste. 1851. *Systeme de Politique Positive*. Translation: *System of Positive Polity*. Reprint, New York: Burt Franklin, n.d.

Comte, August. 1855. *The Positive Philosophy of Auguste Comte*, freely translated and condensed by Harriet Martineau (New York: Blanchard).

Du Bois-Reymond, Emil. 1872. "The Limits of our Knowledge of Nature" (An Address delivered at the 45th Congress of German Naturalists and Physicians at Leipzig). Translated from the German by J. Fitzgerald, *Popular Science Monthly*, vol. 5 (1874), pp. 17-32.

Einstein, A., B. Podolsky and N. Rosen. 1935. "Can Quantum-Mechanical Description of Physical Reality be Considered Complete? *Physical Review*, series 2, vol. 47, pp. 777-80. Reprinted in Toulmin (1970).

Fine, Arthur. 1979. "Einstein's Critique of Quantum Theory: The Roots and Significance of EPR." Preprint, University of Illinois at Chicago Circle, Philosophy Department.

Frank, Phillip. 1941. *Modern Science and its Philosophy* (Cambridge, MA: Harvard University Press).

Frank, Phillip. 1947. *Einstein, His Life and Times* (New York: Knopf).

Heilbron, J. L. 1982. *Elements of Early Modern Physics* (Berkeley: University of California Press).

Hervivel, J. 1966. "Aspects of French Theoretical Physics in the Nineteenth Century," *British Journal for the History of Science*, vol. 3, pp. 109-32.

Holton, Gerald. 1968. "Mach, Einstein, and the Search for Reality," Daedalus, vol. 97, pp. 636-73. Reprinted in his *Thematic Origins of Scientific Thought* (Cambridge, MA: Harvard University Press, 1973), Chapter 8.

Laudan, Larry. 1971. "Towards a Reassessment of Comte's 'Methode Positive'," *Philosophy of Science*, vol. 38, pp. 35-53.

Mach, Ernst. 1882. "Die ökonomische Natur der physikalischen Forschung," *Almanach der kaiserlichen Akademie der Wissenschaften, Wien*, vol. 32, pp. 293-319. English translation by T. J. McCormack in his *Popular Scientific Lectures*, 5th edition (La Salle, Ill.: Open Court Pub. Co., 1943), pp. 186-213.

Malley, Marjorie. 1979. "The Discovery of Atomic Transmutation: Scientific Styles and Philosophies in France and Britain," *Isis*, vol. 70, pp. 213-223.

Maxwell, J. C. 1876. "On the Equilibrium of Heterogeneous Substances," *Proceedings of the Cambridge Philosophical Society*, vol. 2, pp. 427-30; reprinted in *Philosophical Magazine*, vol. 16 (1908), pp. 818-24.

Müller, Erwin. 1956. "Resolution of the Atomic Structure of a Metal Surface by the Field Ion Microscope," *Journal of Applied Physics*, vol. 27, pp. 474-7.

Nye, Mary Jo. 1972. *Molecular Reality: A Perspective on the Scientific Work of Jean Perrin* (New York: American Elsevier).

Nyhof, John. 1981. *Instrumentalism and Beyond*. Ph.D. Thesis, University of Otago, Dunedin, New Zealand.

O'Keefe, John A. and Urey, Harold C. 1977. "The Deficiency of Siderophile Elements in the Moon," *Philosophical Transactions of the Royal Society of London*, vol. A285, pp. 569-75.

Popper, Karl. 1976. *Unended Quest: An Intellectual Autobiography* (La Salle, ILL: Open Court).

Popper, Karl. 1978. "Natural Selection and the Emergence of Mind," *Dialectica*, vol. 32 pp. 339-55.

Popper, Karl. 1980. "Evolution," *New Scientist*, vol. 87, p. 611.

Toulmin, Stephen, ed. 1970. *Physical Reality* (New York: Harper & Row). (Translations of the 1909-10 papers of Planck and Mach; reprint of Einstein et al., 1935.)

Williams, Mary. 1982. "The Importance of Prediction Testing in Evolutionary Biology," *Erkenntnis*, vol. 17, pp. 291-306.

Appendix A

(Comte 1830, Tome I, p. 4)

Enfin, dans l'état positif, l'esprit humain reconnaissant l'impossibilité d'obtenir des notions absolues, renonce à chercher l'origine et la destination de l'univers, et à connaître les causes intimes des phénomènes, pour s'attacher uniquement à découvrir, par l'usage bien combiné du raisonnement et de l'observation, leurs lois effectives, c'est-à-dire leurs relations invariables de succession et de similitude. L'explication des faits, réduite alors à ses termes réels, n'est plus désormais que la liaison établie entre les divers phénomènes particuliers et quelques faits généraux, dont les progrès de la science tendent de plus en plus à diminuer le nombre.

Appendix B

(Tome I, p. 13)

Quant à déterminer ce que sont en elles-mêmes cette attraction et cette pesanteur, quelles en sont les causes, ce sont des questions que nous regardons tous comme insolubles, qui ne sont plus du domaine de la philosophie positive, et que nous abandonnons avec raison à l'imagination des théologiens, ou aux subtilités des métaphysiciens.

Appendix C

(Tome I, pp. 31-32)

En effet, lorsqu'il s'agit, non seulement de savoir ce que c'est que la méthode positive, mais d'en avoir une connaissance assez nette et assez profonde pour en pouvoir faire un usage effectif, c'est en action qu'il faut la considérer; ce sont les diverse grandes applications déjà vérifiées que l'esprit humain en a faites qu'il convient d'étudier. En un mot, ce n'est évidemment que par l'examen philosophique des sciences qu'il est possible d'y parvenir. La méthode n'est pas susceptible d'être étudiée séparément des recherches où elle est employée; ou, du moins, ce n'est là qu'une étude morte, incapable de féconder l'esprit qui s'y livre.

Appendix D

(Tome II, pp. 53-58)

... la diminution de la densité des différentes couches atmosphériques à mesure qu'on s'élève est trop considérable, et d'ailleurs trop intimement liée à la notion même d'atmosphère, pour qu'une telle solution puisse être envisagée comme vraiment rationnelle. Or c'est là ce qui fait la difficulté, jusqu'ici insurmontable, de cette importante recherche. Car il résulte de cette constitution nécessaire de l'atmosphère, non pas une réfraction unique, mais une suite infinie de petites réfractions toutes inégales et croissantes à mesure que la lumière pénètre dans une couche plus dense, en sorte que sa route, au lieu d'être simplement rectiligne, forme une courbe extrêmement compliquée, dont il faudrait connaître le nature pour calculer, par sa dernière tangente comparée à la première, la véritable déviation totale. La détermination de cette courbe deviendrait un problème purement géométrique, d'ailleurs plus ou moins difficile à résoudre, si la loi relative à la variation de la densité des couches atmosphériques pouvait être une fois exactement obtenue; ce qui, en réalité, doit être jugé impossible lorsqu'on veut tenir compte de toutes les causes essentielles.

.... Il faut donc renoncer, au moins dans l'état présent de la science, et probablement aussi pour jamais, à établir d'une manière purement rationnelle une vraie théorie des réfractions astronomiques ... Ainsi, la conclusion pratique de cet ensemble de considérations est qu'il faut, autant que possible, éviter d'observer très près de l'horizon, à cause de la trop grande incertitude des réfractions correspondantes ...

Appendix E

(Tome II, pp. 194-6)

... regarder témérairement une telle extension comme aussi certaine que la gravitation intérieure de notre monde, c'est, à mon avis, altérer autant que possible la nature de nos vraies connaissances, en confondant ce qu'il y a de véritablement positif avec ce qui sera peut-être toujours essentiellement conjectural. En procédant ainsi, on obéit encore, à son insu, à cette tendance métaphysique vers les connaissances absolues, dont l'esprit humain a eu tant de peine à s'affranchir. Sur quoi est fondée la réalité de la gravitation newtonienne? Uniquement sans doute sur sa relation avec les phénomènes, à défaut de laquelle ce ne serait qu'un admirable jeu d'esprit. Or, dans le considération de l'*univers*, il n'y a pas encore de phénomènes exactement observés et mesurés, ... Je crois donc devoir maintenir, en mécanique céleste, comme je l'ai déjà fait en géomètre céleste, la séparation tranchée que je me suis efforcé de rendre sensible, entre la notion de monde et celle d'univers, et la restriction fondamentale que j'ai tâche d'établir, pour nos études vraiment positives, à la seule considération des phénomènes intérieurs de notre système solaire.

Appendix F

(Tome II, pp. 1-5)

L'astronomie est jusqu'ici la seule branche de la philosophie naturelle dans laquelle l'esprit humain se soit enfin rigoureusement affranchi de toute influence théologique et métaphysique, directe ou indirecte; ce qui rend particulièrement facile de présenter avec netteté son vrai caractère philosophique. Mais, pour se faire une juste idée générale de la nature et de la composition de cette science, il est indispensable, en sortant des définitions vagues qu'on en donne encore habituellement, de commencer par circonscrire avec exactitude le véritable champ des connaissances positives que nous pouvons acquérir à l'égard des astres.

Parmi les trois sens propres à nous faires apercevoir l'existence des corps éloignés, celui de la vue est évidemment le seul qui puisse être employé relativement aux corps célestes: en sorte qu'il ne saurait exister aucune asrtonomie pour des espèces aveugles, quelque intelligentes qu'on voulût d'ailleurs les imaginer; et, pour nous-mêmes, les astres obscurs, qui sont peut-être plus nombreux que les astres visibles, échappent à toute étude réelle, leur existence pouvant tout au plus être soupçonnée par induction. Toute recherche qui n'est point finalement réductible à de simples observations visuelles nous est donc nécessairement interdite au sujet des astres, qui sont ainsi de tous les êtres naturels ceux que nous pouvons connaître sous les rapports les moins variés. Nous concevons la possibilité de déterminer leurs formes, leurs distances, leurs grandeurs et leurs mouvements; tandis que nous ne saurions jamais étudier par aucum moyen leur composition chimique, ou leur structure minéralogique, et, à plus forte raison, la nature des corps organisés qui vivent a leur surface. En un mot, pour employer immédiatement les expressions scientifiques les plus précises, nos connaissances positives par rapport aux astres sont nécessairement limitées à leurs seuls phénomènes géometriques et mécaniques, sans pouvoir nullement embrasser les autres recherches physiques, chimiques, physiologiques, et même sociales, que comportent les êtres accessibles à tous nos divers moyens d'observation.

Il serait certainement téméraire de prétendre fixer avec une précision rigoureuse les bornes nécessaires de nos connaissances dans chaque partie déterminée de la philosophie naturelle ... il n'en est pas moins indispensable, ce me semble, de poser à cet égard des limites générales, pour que l'esprit humain ne se laisse point égarer dans le vague de recherches nécessairement inabordables, sans que cependant il s'interdise celles qui sont vraiment accessibles par des procédés plus ou moins indirects, quelque embarras qu'on doive éprouver à concilier ces deux conditions également fondamentales. Cette conciliation si délicate me paraît essentiellement établie à l'égard des recherches astronomiques par la maxime philosophique ci-dessus énoncée, qui les circonscrit dans les deux seules catégories des phénomènes géométriques et des phénomènes mécaniques ... La détermination des températures est probablement la seule à l'égard de laquelle la limite précédemment établie pourra paraître aujourd'hui trop sévère. Mais, quelques espérances qu'ait pu faire concevoir à ce sujet la création si capitale de la thermologie mathématique par notre immortel Fourier, et spécialement sa belle évaluation de la température de l'espace dans lequel nous circulons, je n'en persiste pas moins à regarder toute notion sur les véritables températures moyennes des différents astres comme devant nécessairement nous être à jamais interdite.

Appendix G

(Tome II, pp. 266-67)

... les phénomènes vraiment astronomiques, c'est-à-dire ceux de l'intérieur de notre monde, régissent évidemment tous nos phénomènes sublunaires, soit physiques, soit chimiques, soit physiologiques, soit même sociaux, comme je l'ai indiqué spécialement dans la dix-neuvième leçon. Mais ici nous trouvons, en sens inverse, que les phénomènes les plus généraux de l'univers ne peuvent, au contraire, exercer aucune influence réelle sur les phénomènes plus particuliers qui s'accomplissent dans l'intérieur de notre système solaire. Cette anomalie philosophique disparaîtra immédiatement pour tous les esprits qui admettront avec moi que ces derniers phénomènes sont les plus étendus auxquels nos recherches positives puissent véritablement atteindre, et que l'étude de l'*univers* doit être désormais radicalement détachée de la vraie philosophie naturelle; maxime, à mon avis, fondamentale, et dont j'espère que la justesse et l'utilité seront d'autant plus senties qu'on examinera plus profondément.

Appendix H

(Tome II, pp. 511-13)

Une autre étude qui me semble devoir être radicalement bannie de l'optique, et même de toute la philosophie naturelle, non comme simplement déplacée, mais comme nécessairement inaccessible, consiste dans le théorie de la coloration des corps. Il serait, sans doute, inutile d'expliquer spécialement à ce sujet que je ne saurais avoir en vue, dans une telle critique, l'admirable série d'expériences de Newton et de ses successeurs sur la décomposition de la lumière, qui ont constitue irrévocablement une notion fondamentale, commune à toutes les parties de l'optique. Je veux parler des efforts, nécessairement illusoires, qu'on a si souvent tentés pour expliquer, soit par le systeme émissif, soit par le système vibratoire, le phénomène primordial, évidemment inexplicable, de la couleur élémentaire propre à chaque substance. Ces tentatives irrationnelles sont, à mon avis, des témoignages irrécusables et directs de la fâcheuse influence qu'exerce encore, sur nos intelligences à demi positives, l'antique esprit de la philosophie, essentiellement caractérisé par la tendance aux notions absolues. Il faut que notre raison naturelle soit aujourd'hui bien obscurcie par la longue habitude de ces conceptions vagues et arbitraires que j'ai si souvent signalées, pour que nous puissons envisager, comme une véritable explication de la couleur propre à tel corps, la prétendue faculté de réfléchir ou de transmettre exclusivement tel genre de rayons, ou celle, non moins inintelligible, d'exciter tel ordre de vibrations éthérées, en vertu de telle disposition chimérique des molécules, beaucoup plus difficile à concevoir que le fait primitif lui-même. Les explications placées par l'admirable Molière dans la bouche de ses docteurs métaphysiciens ne sont pas, au fond, plus ridicules. N'est-il-pas déplorable que le véritable esprit scientifique soit encore assez peu developpé, pour qu'on soit obligé de formuler expressément de telles remarques? Personne n'entreprend plus aujourd'hui d'expliquer la pesanteur spécifique particuliere à chaque substance ou à chaque structure. Pourquoi en serait-il autrement quant à la couleur spécifique, dont la notion n'est pas, sans doute, moins primordiale? Cette seconde recherche n'est-elle point, par sa nature, tout aussi métaphysique que l'autre?

Que la considération des couleurs soit, en physiologie, d'une importance capitale pour la théorie de la vision; que, de même, le système de coloration puisse devenir, en histoire naturelle, un moyen utile de classification: cela est évidemment incontestable, et je serais bien mal compris si l'on pouvait penser que je prétends condamner de telles études, ou d'autres tout aussi positives. Mais, en optique, la vraie théorie des couleurs doit se réduire à perfectionner l'analyse fondamentale de la lumière, de manière à apprécier l'influence de la structure, ou de telle autre circonstance générale, même accidentelle ou fugitive, sur la couleur transmise ou réfléchie, sans jamais s'engager d'ailleurs dans la recherche illusoire des causes premières de la coloration spécifique: le champ d'études ainsi circonscrit offre, certainement, encore une assez vaste carrière à l'activité des physiciens.

Appendix I

(Tome III, pp. 28-29)

Toute tentative de faire rentrer les questions chimiques dans le domaine des doctrines mathématiques doit être réputée jusqu'ici, et sans doute à jamais, profondément irrationnelle, comme étant antipathique à la nature des phénomènes: elle ne pourrait découler que d'hypothèses vagues et radicalement arbitraires sur la constitution intime des corps, ainsi que j'ai eu occasion de l'indiquer dans les prolégomènes de cet ouvrage.

Appendix J

(Tome II, p. 18)

Aucune partie de la philosophie naturelle ne peut donc manifester avec plus de force la vérité de cet axiome fondamental: *Toute science a pour but la prévoyance*, qui distingue la science réelle de la simple érudition, bornée à raconter les événements accomplis, sans aucune vue d'avenir.

(Tome III, p. 4)

En introduisant, dans des actes chimiques déjà bien explorés, quelques modifications déterminées, même légères et peu nombreuses, il est très rarement possible de prédire avec justesse les changements qu'elles doivent produire: et neanmoins, sans cette indispensable condition, comme je l'ai si fréquemment établi dans ce traité, il n'existe point, à proprement parler, de *science*; il y a seulement *érudition*, quelles que puissent être l'importance et la multiplicité des faits recueillis. Penser autrement, c'est prendre une carrière pour un édifice.

B.
LAWS

TRANSCENDENT LAWS AND EMPIRICAL PROCEDURES

James H. Fetzer

My purpose here is to provide a sketch in broad and general language of the basic reasons for the (unavoidable) uncertainty of scientific knowledge. The proper relationship between the history of science and the philosophy of science, I think, comes to this: that the *aim* of science can only be ascertained by investigating the history of science, while the *methods* of science are correctly established on the basis of philosophical considerations. The difference thus displayed reflects one difference between "means" and "ends," for although "ends" are pragmatically contrived, the most appropriate "means" for their attainment need not be the same as those which we happen to employ. The combination of a normative theory of "means" with a descriptive theory of "ends," no doubt, may seem perplexing to some; but surely there are no built-in guarantees that the methods employed by many or by most or by all who call themselves *members of the scientific community* are necessarily the most effective, the most reliable, or the most efficient "means" to achieve a given goal—even if that goal coincides with the aim of science itself.

The suggestion that the aim of science can only be ascertained by investigating the history of science, however, does not mean that this is a matter beyond debate about which historians could not possibly disagree; for in fact it clearly qualifies as an issue involving *questions of interpretation*, where the most plausible, logically consistent, and theoretically illuminating analysis requires identification and historical defense. Thus, within this forum, I shall assume that an investigation of the history of science offers evidence sufficient to sustain the inference that science does have its own distinctive aim, namely: the discovery of general principles by means of which the phenomena of experience may be systematically subjected to explanation and prediction, where these "general principles" possess the form of scientific theories and of natural laws. Indeed, scientific theories themselves are appropriately entertained as sets of laws which apply to a common domain, thereby permitting the concise depiction of science as aiming at the discovery of natural laws. Any other activity could be perfectly worthwhile, but it could not be science.

The aim of discovering natural laws, of course, might be accepted as necessary without being sufficient as a historical portrayal; for the methods of scientific discovery are ordinarily restricted to the results of inter-subjectively accessible observation and experimentation and to the inferences those results could sustain. The controversies that rage over the most appropriate "means" for attaining such an "end" thus commonly presuppose the availability of *experiential findings* in the form of a (perhaps quite large) finite set of logically consistent singular sentences e, let us say, where the primary bones of contention are possible *"principles of inference"* purporting to codify the precise conditions under which such a set of sentences e provides "evidential support" to various members of a set of alternative *general hypotheses* h^1, h^2, and so on. The history of science not only determines the goal of scientific inquiry, therefore, but also defines the global features of its methods themselves to the extent to which the discovery of natural laws is supposed to rely upon the employment of empirical procedures.

It cannot be denied that the availability of any specific evidence e itself may depend upon historical conditions, since the performance of relevant experiments and the conduct of relevant observations frequently involves technical apparatus, such as telescopes and cyclotrons, which permit observations and experiments that might otherwise remain inaccessible. The particular circumstances that impose *practical limitations* upon the scope of scientific discovery, moreover, are not generally confined to the category of technological innovations, for the preparation and training of qualified personnel and the allocation and assignment of necessary funding obviously create their own distinctive restraints. Perhaps more interesting to contemplate, therefore, from a philosophical point of view, are issues concerning the existence and character of *inherent limitations* with respect to the scope of scientific knowledge, restrictions or constraints which no technological innovations, qualified personnel, or necessary funding could possibly overcome, because they arise from a tension between the aims of science and the global features of its methods.

The ultimate source of this tension, of course, is the nature of natural laws. Most theoreticians would agree, I believe, that laws involve relations between properties, although some hold out for classes instead. Classes, as we all know, are completely extensional entities, since two classes are said to be the same so long as their members are the same, a conception which combines with truth-functional commitments to yield an account of "natural laws" as describing *constant conjunctions* and *relative frequencies* across classes. Consider, however, the following columns of property (or class) designators:

(I) R A

red	round
wooden	cuckoo-clock
gold	melting point of 1063° C.
polonium[218]	half-life of 3.05 min.

No doubt, there is some relative frequency with which things-that-are-red are things-that-are-round, with which things-that-are-wooden are things-that-are-cuckoo-clocks, and so forth, as features of the history of the world. Equally clearly, no doubt, only some *but not all* of these true extensional distributions would be seriously supposed to represent "natural laws."

From an extensional perspective, this circumstance poses a delicate predicament, since the separation of distribution descriptions that are *true and laws* from those that are *true but not* seems to presuppose some non-extensional principle of selection, which undermines the integrity of this approach. Some have thus imagined that this difference is either a "question of context" or a "matter of attitude," embracing pragmatical contrivances to avoid intensional entanglements. But their concrete proposals have not been especially reassuring, since attitudinal advocates incline toward the criterion that true extensional distributions are laws when they are *regarded* as being laws, as though that should explain why those claims are properly regarded as laws; while the contextual adherents recommend instead the standard that those are laws which can be *derived* from scientific theories, as if theories were not sets of laws themselves. In order to preserve this extensional conception, in other words, serious thinkers have felt compelled to appeal to blatantly circular maneuvers and to overtly question-begging stratagems.

Within an intensional framework, by comparison, things are not so desperate, for there appear to be ample resources for contending with these differences without resorting to *ad hoc* principles. Even if 1% of all things-that-are-wood were things-that-are-cuckoo-clocks, or if 99% of all things-that-are-cuckoo-clocks were things-that-are-wood, those extensional distributions, as true descriptions of the world's history, would not *therefore* qualify as "natural laws." For there are processes and procedures, such as the production of plastic cuckoo-clocks, or of metal cuckoo-clocks, or the passage of legislation prohibiting the use of wood in their construction, by virtue of which, in principle, these extensional relations would be subject to change—even if during the world's history common practice should prevail and those relations should remain invariant! In the case of gold and of polonium[218], by contrast, matters are quite different, for there appear to be *no* processes or procedures by means of which the melting point of 1063° C. and the half-life of 3.05 min. could be taken away from things of those kinds—were we disposed to try!

The intensional conception thus displays the sense in which natural laws are correctly entertained as *negative existential propositions*; for laws hold as intensional relations between a reference property R, let us say, and some attribute property A, where a lawful connection obtains between those properties just in case there is no process or procedure, either natural or contrived, by means of which something possessing the property R could lose the property A without also losing the property R, where the possession of A by something that is R remains, nevertheless, *logically contingent*, i.e., the predicate 'R' entails neither the predicate 'A' nor its negation '$\sim A$', within the relevant language. Consequently, strictly speaking, the logical form of lawlike sentences, i.e., of sentences that would be natural laws if they happen to be true, is that of unrestrictedly general, logically contingent, subjunctive conditionals attributing *permanent properties* A to everything possessing reference properties R. The subjunctive conditionality of natural laws thus reflects their ontological character rather than any pragmatic circumstances.

It should not be overlooked, therefore, that the attributes of (merely) extensional distributions are not properties that no members of their reference classes could be without: things-that-are-round might or might not be things-that-are-red, things-that-are-wood might or might not be things-that-are-cuckoo-clocks, and so on. As a result, these properties are appropriately entertained as *transient attributes* of the members of their corresponding reference classes. But it is important to observe as well that natural laws invariably qualify as *distributive* as opposed to *collective* generalizations, i.e., they attribute a permanent property A (such as a melting point of $1063°$ C. or a half-life of 3.05 min.) to every thing possessing the reference property R (such as being gold or being polonium[218]), instead of summarizing the relative frequency with which those attributes occur within those populations. Indeed, it may be said, as a perfectly general principle, that any extensional distribution describing the relative frequency for an attribute A within a reference class R cannot possibly qualify as a natural law!

That attribute A occurs in constant conjunction with a reference property R, of course, thus constitutes a necessary but not sufficient condition for a lawful connection to obtain between them, because A might happen to be a transient property possessed by every member of a reference class R in common (as would be the case if every cuckoo-clock *were* made of wood). The distributive character of natural laws stands in sharp contrast with means and medians and modes as the following columns of descriptive predicates are meant to display:

(II)	R	A
	gifted children	130 median I.Q.
	white Anglo-Saxons	usually Protestant
	BMW's	average $15,000.
	New College	mean 600 SAT's
	students	

Even if gifted children (identified by their verbal abilities) happen to have a median I.Q. of 130, that attribute could not possibly be a permanent property of the members of that reference class—unless *every* gifted child happened to have that same I.Q. in common; even if white Anglo-Saxons usually are Protestants, that attribute could not possibly be a permanent property of the members of that reference class—unless *every* white Anglo-Saxon happened to be a Protestant; and so on. For permanent properties are properties no member of a reference class could be without and thus permit of no exceptions.

The apparent tension between the aims of science and the global features of its methods to which I have alluded above, therefore, emerges clearly from this point of view. For science aims at the discovery of natural laws on the basis of experiential findings and the inferences they sustain, yet there are no apparent "principles of inference" relative to which any logically contingent, unrestrictedly general subjunctive conditionals h attributing permanent properties to every member of appropriate reference classes could be securely established, in principle, on the basis of (even quite large) finite sets of logically consistent singular sentences e. The difficulties involved here, I should emphasize, are not (merely) those of warranting inferences from finite samples to infinite populations, but those of warranting inferences from evidence describing (segments of) the world's *actual history* to hypotheses concerning the *possible histories* it could display, under differing initial conditions! For the logical force of natural laws as negative existential propositions not only entails that it is not the case that anything *satisfies* the corresponding descriptions "R & $\sim A$" during the world's history (an "ordinary" inference) but also that it is not the case that anything *could satisfy* those descriptions during any such history (an "extraordinary" inference, indeed)!

In order to appreciate the inherent limitations with respect to the certainty of scientific knowledge thereby generated, let us consider (at least) three alternative conceptions of "principles of inference" which potentially might serve to "bridge the chasm" between evidence and hypothesis. According to these views, science proceeds in stages along roughly the following lines:

(III)	INDUCTIVISM	DEDUCTIVISM	RETRODUCTIVISM
	Observation	Conjectures	Puzzlement
	Classification	Derivations	Speculation
	Generalization	Experimentation	Adaptation
	Explanation	Elimination	Acceptance

Alternative Conceptions of Scientific Procedure

These are not the only available possibilities, of course, since other views, including "The Bayesian Way" with its infinite contours, have vocal advocates too, but as non-exhaustive illustrations perhaps these three will do. Indeed, Inductivism, Deductivism, and Retroductivism have been historically important and philosophically influential theoretical formulations which surely deserve consideration as significant conceptions of scientific procedure. Whether or not approaches of any of these kinds should or should not be taken seriously, furthermore, depends far less upon their specific details than upon the sort of "principles of inference" they propose to attain the objective of inquiry.

From this point of view, therefore, the inductivist conception of scientific procedure as a pattern of Observation, Classification, Generalization, and Explanation assumes importance only in relation to the basic "principles of inference" which establish its foundation; thus, according to one such approach, the aim of science should be pursued by the method of enumerative induction:

(IV) From "m/n observed R's are A's," infer to "m/n R's are A's";

provided that a large number of R's have been observed over a wide variety of conditions, where all such inferences are subject to revision with the accumulation of additional evidence. If natural laws are subjunctively conditional while relative frequencies are extensional distributions, however, methods of this kind afford no criterion for the identification of those descriptions which are true-and-laws as opposed to those which are true-but-not. Even the wide variety requirement does not help: a large number of cuckoo-clocks have been observed in many different locations at many different times under quite diverse circumstances, yet the corresponding distribution establishes no law. The pattern of Observation, Classification, and Generalization by this method thus appears incapable of satisfying conditions required for discovering laws.

The deductivist conception of scientific procedure promises to do better, albeit in a negative direction, for the pattern of Conjecture, Derivation, Experimentation, and Elimination at least allows for the rejection of genuinely lawlike claims through the employment of the method of eliminative induction:

(V) From "hypothesis h entails e" and "not-e," infer to "h is false";

provided that *rejection* is not mistaken for *disproof*, since the evidence upon which such inferences are based may itself turn out to have been at fault, as a function of background assumptions, auxiliary hypotheses, and other sources of potential error. Since natural laws as subjunctive generalizations entail corresponding extensional distributions in the form of constant conjunctions, the discovery of even one R that is not an A is sufficient to sustain the rejection of the corresponding lawlike claim. Yet so long as explanations are arguments, their assertion as adequate entails their acceptance as true: insofar as natural laws are necessary for adequate explanations, science cannot succeed without procedures for acceptance as well as principles of rejection.

The deductivist conception nevertheless signals an enormous improvement by comparison with its inductivist counterpart, since it encourages attempts to arrange this world's history so that it should include sets of events of evidential relevance for testing alternative hypotheses and scientific theories. Indeed, it thus becomes apparent that the aim of science clearly requires recognition of the difference between *confirming* extensional distributions and *testing* lawlike claims, for the evidence relevant to lawlike claims consists of repeated attempts to refute them! The retroductivist conception thus affords an appropriate "complement" to eliminative induction, at least so long as the stages of Puzzlement, Speculation, Adaptation, and Acceptance are entertained within the framework of "inference to the best explanation," which can be captured by likelihood measures of evidential support such as follows:

(VI) From "the nomic expectability of e, given h, equals r," infer to
        "the measure of evidential support for h, given e, equals r";

with the understanding that a large number of trials have been conducted over a wide variety of conditions, where these measures too are subject to modification with the acquisition of still more evidence. And, although standards of this kind are properly regarded as measures of *preferability* as opposed to measures of *acceptability* (when applied to alternative hypotheses), there are reasons to suspect that, with appropriate qualifications, an approach of this type lends itself to the formulation of (plausible) principles of acceptance.

The retroductivist conception, of course, is not without its own distinctive difficulties, insofar as every consistent theory entailing an hypothesis h satisfying the specified conditions receives a corresponding measure of evidential support from the evidence e, a problem partially offset by employing explanatory relevance and irrelevance relations as a foundation for determining evidential relevance and irrelevance relations (for scientific theories). Not the least of the problems I have left unmentioned, moreover, is the need for *nature's cooperation* in the pursuit of natural laws; for where there are no samples at all, or only small samples, or even large (but skewed) samples, the available evidence may warrant either a faulty inference or no inference at all. This difficulty becomes acute with respect to probabilistic properties, insofar as any relative frequency within any finite sequence is logically compatible with any probabilistic attribute, even though deviations from generating probabilities become increasingly improbable as the length of such a trial sequence increases without bound. Thus, even over infinite sequences it remains not merely logically but indeed physically possible that limiting frequencies for various outcomes may diverge from corresponding probabilities.

From an epistemic point of view, perhaps the most intriguing consequence consists in the result that a physical world whose composition includes (some) probabilistic properties might be historically indistinguishable from a world whose composition includes *no* probabilistic properties—either because there are no appropriate trials, or too few appropriate trials, or enough appropriate trials which, by chance, happened to yield unrepresentative frequencies. It should be emphasized, therefore, that even if a world were *indeterministic* in its character (by instantiating probabilistic as well as non-probabilistic properties), the history of that world could turn out to be indistinguishable from the history of a *deterministic* world insofar as they might both display exactly the same relative frequencies and constant conjunctions, where their differences were concealed "by chance"! Thus, if some of the world's properties are probabilistic, not only may the same laws generate different world-histories, but the same world-histories may be generated by different laws—under identical initial conditions! Indeed, the most important implication attending these considerations deserves explicit recognition, since even if our available evidence could describe the world's entire history, fundamental aspects of its ontological structure might remain undiscovered, nevertheless. The tension between the aim of science and the global features of its methods therefore cannot be overcome: the uncertainty of scientific knowledge is a direct effect of the nature of natural laws.[1]

NOTES

1. For a more technical elaboration and formal defense of the themes developed here, especially in relation to probabilistic properties and corresponding laws, see James H. Fetzer, *Scientific Knowledge* (Dordrecht, Holland: D. Reidel, 1981).

CONCEPTIONS OF SCIENTIFIC LAW AND PROGRESS IN SCIENCE

Davis Baird

There are at least three constraints determining the limits of science: the way nature is; our cognitive ability to find out about nature; and what counts as a *scientific* product of our attempts to learn about nature. In this paper I address the third constraint.

Among the most important products of our attempts to understand nature are scientific laws. Many argue that scientific laws must be predictive. Isolated or wildly variable phenomena are not subsumable under such laws then. They consequently fall outside the limits of science. I want to block this sort of limiting argument.

I argue that problems come up in science which are best attacked with scientific laws which are assumed to be purely descriptive. Scientific laws can be simply descriptive, they need not be explanatory or predictive or something else as well. I do not argue that laws are only descriptive; no doubt problems come up in science which are best attacked with other conceptions of scientific law.

The bulk of my argument is drawn from the work of Karl Pearson (1857–1936). I focus on two aspects of Pearson's work: his philosophical reflections on the nature of scientific laws, and his important work on what he called the "Generalized Normal Law." Pearson's "Generalized Normal Law" was the first step freeing statisticians from the omnipresent Normal curve of errors—or "bell-shaped" probability curve—towards our modern measure-theoretic concept of probabilities. Pearson held firmly and explicitly the view that scientific laws are only descriptions of our past experience: they explain nothing. Pearson's conception of the nature of scientific laws as purely descriptive is discredited today, but I show that it was the key to his invention of the Generalized Normal Law.

Can Scientific Laws Be Purely Descriptive?

It is almost a platitude that laws cannot be purely descriptive. Laws must be distinguished from accidental generalizations. Different methods to make the distinction have been tried. Hempel emphasizes the role of laws in explanation (Hempel, 1965, p. 56). Goodman emphasizes the role of laws for prediction (Goodman, 1955, p. 26). According to either line of thought, explanation or prediction are essential to laws; description comes by the way.

Pearson denies that laws explain facts:

> The discussion of the previous chapter has led us to see that law in the scientific sense only describes in mental shorthand the sequence of our perceptions. It does not explain why these perceptions have a certain order, nor why that order repeats itself. (Pearson, 1892, p. 113).

Pearson emphasizes this point many times throughout the *Grammar*. He concludes after a brief discussion of the development of laws on celestial motion:

> The law of gravitation is a brief description of *how* every particle of matter in the universe is altering its motion with reference to every other particle. It does not tell us *why* the earth describes a certain curve around the sun. (Pearson, 1892, p. 99).

One might argue that Pearson employs a different sense of explanation from that of Hempel. Thus Pearson's negative claim that laws do not explain need not conflict with Hempel's claims to the contrary. Pearson's positive claim that laws are purely descriptive is, however, more difficult to reconcile with Hempel. Few philosophers would be happy to explain the Earth's distance from the sun by reference to Bode's Law and the fact that the Earth is the third planet of the eight planets for which Bode's Law is accurate.

Pearson is somewhat more generous with the predictive use of laws:

> Science for the past is a description, for the future a belief; . . . If the reader has once fully grasped that science is an intellectual *resume* of past experience and a mental balancing of the probability of future experience, he will be in no danger of contrasting the 'mechanical explanation' of science with the 'intellectual description' of mythology. (Pearson, 1892, p. 113).

Pearson understands science broadly as a two-part enterprise: in the first part the scientist's task is to find accurate economical descriptions of how things have been going—that is, the scientist tries to find (Karl Pearson-style) laws; in the second part the scientist's task is to estimate, with the help of probability relations, how likely the patterns described by laws are to repeat themselves. Laws *per se* are not used for prediction; laws in conjunction with some probability magic are used for prediction. Pearson turns Hempel and company on their heads: laws are essentially descriptive; as a matter of fact they may be used for prediction and understood correctly for explanation.

Supporters of the explanatory view of scientific laws frequently cite the existence of "accidental generalizations" as conclusive evidence for the explanatory view. Hempel points out that "All bodies of pure gold have a mass less than 100,000 kilograms" is not a law, even though it is certainly not refuted and very possibly is true. Here is a descriptive statement which is very likely true, but, says Hempel, is not a law. Consequently there must be more to being a law than simply descriptive accuracy. The problem is to distinguish statements that *could be* laws from other descriptively accurate accidental generalizations.

Pearson would object to this line of argument on at least three grounds. In the first place he would deny that the statement, "All bodies of gold have a mass of less than 100,000 kilograms," is not a law. For Pearson it is simply not a very good law. Laws, for Pearson, aim to describe — as briefly as possible — sense experience. This statement is descriptive; it is not, however, particularly economical. Our reluctance to call it a law simply reflects the fact that it summarizes only a small bit of experience in a whole sentence.

Pearson would also point out that the distinction between accidental and law-like has not been made out. Pearson does not argue this explicitly; my remarks are based on an interpretation of what he does say. In 1898 Pearson discovered what is now known as the problems of "spurious correlation" or "Simpson's paradox" (Pearson, 1898). This problem is now taken to demonstrate that there is a fundamental difference between high correlation and causation. Pearson drew no such inference. For Pearson, we have no concept of causation other than that of high correlation:

No phenomena are causal; all phenomena are contingent, and the problem before us is to measure this contingency, which we have seen lies between the zero of independence and the unity of causation. (Pearson, 1911, p. 194).

By "the zero of independence and the unity of causation" Pearson refers to his scale of statistical correlation. The only way I see to understand Pearson's view of correlation and his interpretation of the problem of spurious correlation is to admit that there is no difference between law-like regularities and accidental regularities. All correlations are equally good; there is no difference to be drawn between those that mirror the "nomological structure" of the world and those that are just accidental. For Pearson the only difference is that of economy and accuracy.

Pearson's third objection contrasts what I call the "prescribed usage" and the "methodological" interpretations of my question, "Can laws be purely descriptive?":

Prescribed Usage: What statements ought we call scientific laws? Ought we call purely descriptive statements laws or must we include an explanatory or predictive element for laws?

Methodological: Should a scientist aim to discover (or invent) purely descriptive laws or must a scientist aim at something more, e.g. predictive power?

Perhaps after all is said and done it will be necessary to distinguish between laws and accidental generalizations. But for the practicing scientist all is never said and done. Consequently, for Pearson, the prescribed usage interpretation is secondary to the methodological interpretation. One might argue that the reverse is true: only once we understand what laws are in essence shall we be able to identify how to seek such beasts. Indeed, this is a common pattern of argumentation: first identify the essential features of some problem by way of conceptual analysis, then distinguish the "important questions" from the "pseudoquestions." For Pearson the procedure is the reverse: one must first scrutinize the problems facing the scientist and the methods used for coping with these problems. Only then can one reach important conclusions about the scientific enterprise.

In particular, for Pearson an important question is, "What is a good classification of the facts and what are the facts so classified?" A bad question is, "What is the difference between laws and accidental generalizations?" Pearson tells us that epoch-makers in science are those who provide us with new means to appreciate and analyze the phenomena before us:

The scientific discoverer is one who by exercise of the imagination widens our appreciation of phenomena or improves our methods of analysing our appreciations. (Pearson, 1919, p. x).

The passage is difficult to interpret without a clear understanding of what Pearson means by "appreciation of pehnomena." I think a hint of what he has in mind comes from the previous sentence:

All scientific truth is relative — relative to the assemblage of facts before us at the time — but what is, perhaps, more important relative to the facts as they are conceived by our current mentality on the basis of the perceptions of them provided by the instruments of measurement and observation of our own day. No scientific truth is absolute, it is relative to our appreciation of phenomena, and to our methods of classifying our appreciations. (Pearson, 1919, p. x).

Pearson is not saying "anything goes, it is all a matter of how you look at it." For Pearson the rock bottom of all knowledge is based in sense data. For an individual these are unquestionable. It is what we do with out sense data that Pearson has in mind when he speaks of our appreciation of the phenomena. This is the important issue of understanding scientific thought. This is why Pearson insists that scientific law is descriptive.

Pearson's argument rests on several premises whose truth would likely not be granted today. Is there a "rock bottom" of sense date, or just plain data for that matter, out of which we construct our scientific appreciations? Is avoiding accidental generalizations not a real worry for the practicing scientist? Can we determine what problems face the scientist without some, perhaps incomplete, notion of what laws essentially are? I do not think we need accept Pearson's conclusion that laws are descriptive. However, I do not think Pearson's point of view is wholly wrong either. There are occasions when one difficulty which faces the scientist is to find a good classification of phenomena. On such occasions aiming for purely descriptive laws is more beneficial than aiming for explanatory laws. Pearson's invention of the Generalized Normal Law is a case in point.

THE NORMAL LAW

In order to understand Pearson's Generalized Normal Law we must briefly consider the (ungeneralized) Normal law. One of the central problems which faced the 19th Century scientist was the problem of how to handle systematic variation. This problem first yielded a solution with Gauss's Theory of Error for observations of the heavens. Gauss's solution, to represent variations with what we now call the Normal Probability curve, was generalized throughout the 1800's and applied to most other problems where systematic variation was found.

Indeed arguments were presented to show that Gauss's solution had to generalize. The following argument had considerable currency during the 1800's. We assume that there are many causes for error. Each of these causes, taken by itself, is negligible. Noticeable errors result from the combined action of many of these individual small causes. But the small causes of error can "push" a measurement either greater than or less than the true value. We suppose that the probability of either sort of "push" from any individual error is the same, i.e., $1/2$. So we can think of the interaction of all these "small causes" as the results of flipping a fair coin many times: each head pushes the observed value greater than the true value; each tail pushes the other way. The net effect is captured in the ratio of heads to tails. While on average there are just as many heads as tails in sequences of flips of a fair coin, it is possible for there to be more heads or more tails — resulting in errors. We can calculate the probability of possible sequences of heads and tails with s heads and $N-s$ tails in N flips; this gives a measure of the frequency of a measurement $N/2-s$ units from the true value. As the number of flips grows without bound — as the number of little contributory causes of error grows without bound — these coin-flipping probabilities approach the Normal curve. Hence the Normal curve represents the distribution of errors around the true value.

Such *a priori* arguments combined with the empirical discovery during the mid-1800's that many variable phenomena can be well represented with the Normal curve elevated the Normal curve to the status of the "Normal Law." The Normal Law "governed" cases where variation of a type was manifest — in particular with various size measurements on species. Francis Galton writes in 1889:

> *Order in apparent Chaos:* — I know of scarcely anything so apt to impress the imagination as the wonderful form of cosmic order expressed by the 'Law of Frequency of Error'. The (Normal Frequency) law would have been personified by the Greeks and deified, if they had known of it. It reigns with serenity and in complete

selfeffacement amidst the wildest confusion. The hugher the mob, and the greater the apparent anarchy, the more perfect is its sway. It is the supreme law of unreason. (Galton, 1889, p. 66).

Indeed as late as the 1911, 11th edition of the *Encyclopaedia Britannica* one finds F. Y. Edgeworth quoting in his article on 'Probability" a proof of the Normal law by the physicist John Herschel (Edgeworth, 1911, Vol. 22, p. 391).

John Venn launched a critique of the "Normal Law" as early as 1876. But criticism is not enough. Pearson presented an alternative: the Generalized Normal Law.

THE GENERALIZED NORMAL LAW

Starting in 1894 Pearson published a series of papers with the title, "Mathematical Contributions to the Theory of Evolution." The first two contributions were devoted to extending the techniques available to field biometricians for describing and analyzing data. (The third presents for the first time a rigorously worked out theory of statistical correlation.) The need for new techniques to deal with biological data arose because non-Normal data was found. The biologist W. F. R. Weldon found in measuring Naples crabs that the frequency curve that resulted was "doublehumped"; the measurements did not vary around a single most frequently observed value, but instead had two peaks for two very frequently observed values. The statistician and economist F. Y. Edgeworth, referred to above, found frequency measurements of economic indices whose peak, or most commonly observed value, was significantly pushed to one side or another. In his first contribution Pearson attempted to deal with the non-Normal character of Weldon's observations by presenting techniques for analyzing double-humped data into two Normal components. In his second contribution Pearson developed the Generalized Normal Law. This was an equation with several parameters; by selecting specific values for the parameters, a specific frequency curve was determined. Pearson's equation was capable of determining any Normal curve, as well as many others besides the Normal.

Pearson's first contribution is a testament to the sway of the Normal Law. He justifies his attempt to reduce non-Normal data to Normal components in two ways. In the first place, Normal variation was one criterion identifying a species. Thus it was reasonable to suppose that Weldon's "double-humped measurements" of Naples crabs was evidence that the species was in process either of differentiating or of merging. Pearson's Normal components then could be supposed to represent the two partially connected species. On the other hand, when it was totally implausible to suppose that the non-Normal data was heterogeneous in this way, (i.e., comprised of two distinct populations) Pearson justified his determining Normal components with the following remark:

> Even where the material is really homogeneous but gives an abnormal frequency-curve the amount and direction of the abnormality will be indicated if this frequency-curve can be split up into Normal curves. (Pearson, 1894, p. 2).

Normalcy was the paradigm against which all frequency data were judged. All of this changes with Pearson's second contribution.

Pearson seeks in his second contribution to find a function expressed by several parameters which—by varying the values of the parameters—is sufficient to accurately describe all of the possible sets of frequency data one might find in nature. Pearson's set of frequency curves is now seen to be hopelessly narrow. It must be remembered, however, that Pearson did not have the abstract measure theoretic conception of probability we have today. He had an empirically inadequate Normal Law. Since the Normal Law was reasonable in many cases, Pearson's attempt to generalize it was the obvious thing to do.

Pearson proceeds as follows: First he recalls the limiting relationship between the Normal curve and binomial probabilities for a fair coin: $(S!/s!(S\text{-}s)!)\,(1/2)^s(1/2)^{S-s}$. He argues that asymmetric curves may be found if the binomial parameter is allowed to differ from $1/2$ giving what he calls asymmetric point-binomials: $(S!/s!(S-s)!)P^sq^{S-s}$, where $p + q = 1$. Pearson observes that there is an important relationship between the Normal curve and the symmetic point binomial besides the limiting relationship:

> Hence: this binomial polygon and the normal curve of frequency have a very close relation to each other, of a geometrical nature, which is quite independent of the magnitude of n. In short, their slopes are given by an identical relation. (Pearson, 1895, p. 53).

The relation is:

$$\frac{\text{slope of curve}}{\text{ordinate}} = \frac{dy}{dx}\frac{1}{y} = \frac{2\ \text{abcissa}}{2\ \sigma^2} \quad \text{(Pearson, 1895, p. 54).}$$

This relationship strikes Pearson as important; he endeavors to determine a curve which bears the same relation to the asymmetric point binomial.

Pearson is not entirely satisfied; the curve so determined is limited in range in one direction and indefinitely extended in the other. This consideration of range leads to the Pearson System of probability curves:

Type I.   Limited range in both directions and skewness.
Type II. Limited range and symmetry.
Type III. Limited range in one direction only and skewness.
Type IV. Unlimited range in both directions and skewness.
Type V.   Unlimited range in both directions and symmetry.
          (Pearson, 1895, p. 58).

Pearson finds a formula that is capable, on appropriate substitutions for the free parameters, of realizing any of these forms. All of his curves retain the geometric relationship noted above. The remainder of the paper is devoted to URN models and empirical examples. Pearson's solution, given his understanding, seems to me straightforward and natural.

CONCLUSION

In his second paper Pearson takes an important methodological stand: the aim of a scientific law is to accurately and economically describe phenomena; the test of any proposed law is the degree to which it performs these functions:

> As the great and only true test of the normal curve is: Does it really fit observations and measurements of a symmetrical kind? So the best argument for the generalized probability curve deduced in this paper is that it does fit, and fit surprisingly accurately observations of an asymmetrical character. (Pearson, 1895, p. 43).

While the normal curve could be fit to much frequency data with differences chalked up to sampling error, Pearson abandoned this explanation as 'ad hoc' and proposed his more general system of curves:

> Indeed, there are very few results which have been represented by the normal curve which do not better fit the generalized probability curve, a slight degree of asymmetry being probably characteristic of nearly all groups of measurements. (Pearson, 1895, p. 43).

Of course Pearson could describe data as accurately as is possible by listing the data. The point is to describe economically. The Normal Law was economical; it was not very accurate. By introducing a few more parameters, Pearson greatly improved the accuracy of the available scientific descriptions with only a small cost in economy.

If Pearson had been wedded to the idea that scientific laws must explain as well as describe he should not have invented the Generalized Normal Law. In the first place, differences between frequency data and the best fit Normal curve could always be dismissed with the claim, "bad sample." Secondly, and more importantly, the Normal Law explained where Pearson's Generalized Normal Law simply described. There were widely accepted arguments, such as that presented above, explaining why variations in measurements should vary with a Normal frequency. These arguments linked Normal variation with the identity of a species: non-Normal data did not come from a true species. The *a priori* arguments explained why variations must be Normal and the Normal Law explained the identity of species amidst variation. By advocating the Generalized Normal Law Pearson lost all these explanatory functions; his law could only describe the data more accurately than the Normal Law. Pearson had to advocate descriptive accuracy over explanatory power to see this Generalized Normal Law as an improvement over the Normal Law.

I do not wish to maintain that this one example of Karl Pearson and the Generalized Normal Law shows that laws are descriptive after all. Laws perform many functions; they are descriptive, explanatory and predictive. I do wish to maintain that it is wrong to hold the view that laws have any one or more of these functions to the exclusion of the others. Laws help us solve problems; different problems require a different emphasis on the nature of laws. Pearson's need to develop a means to deal with systematic variation was aided by emphasizing the descriptive aspect of laws. Other needs may well require other emphases. Perhaps future problems will require entirely new and different conceptions of scientific law. We ought not limit the progress of science by adopting some definition of law simply to answer questions about the limits of science.

References

Carnap, Rudolf, 1966. *An Introduction to the Philosophy of Science*, ed by Martin Gardner (New York: Basic Books).

Edgeworth, F. Y., 1911. "Probability" article in the *Encyclopaedia Britannica*, 11th edition, Vol. 22, pp. 376–403.

Galton, Francis, 1889. *Natural Inheritance* (London: Richard Clay and Sons).

Goodman, Nelson, 1955. *Fact, Fiction and Forecast* (Cambridge, MA: Harvard University Press).

Hempel, Carl, 1966. *Philosophy of Natural Science* (Englewood Cliffs: Prentice Hall).

Pearson, Karl, 1892. *The Grammar of Science*, see Pearson, 1957.

Pearson, Karl, 1894. "Contributions to the Mathematical Theory of Evolution," reprinted in Pearson, 1948. Page references are to that volume.

Pearson, Karl, 1898. "Mathematical Contributions to the Theory of Evolution—II. Skew Variation in Homogeneous Material," reprinted in Pearson, 1948. Page references are to that volume.

Pearson, Karl, 1898. "Mathematical Contributions to the Theory of Evolution—IX. On a Form of Spurious Correlation which may Arise when Indices are used in the Measurement of Organs." *Philosophical Transactions of the Royal Society*, A, Vol. 192.

Pearson, Karl, 1919. Introductory essay to: Goring, Charles, *The English Convict* (London: Butterworth).

Pearson, Karl, 1948. *Karl Pearson's Early Statistical Papers* ed. by E. S. Pearson (London: Cambridge University Press).

Pearson, Karl, 1957. *The Grammar of Science* (New York: Meridan Books). This is a reprint of the third 1911 edition. I have cited Pearson, 1892 to indicate that the material referred to appeared in the first 1892 edition. Page references, however, are to the 1957 reprint.

Venn, John, 1962. *The Logic of Chance* (New York: Chelsea Publishing Co.). This is a reprint of the third 1888 edition. Venn's critique of the Normal Law first appears in the second 1876 edition.

CAN SCIENCE KNOW
WHAT'S NECESSARY?

Dennis Temple

INTRODUCTION

Few claims concerning the limits of scientific knowledge have been as durable as Hume's denial that science can discover necessary connections in nature. Hume's well-known arguments on this matter cover the following four points: (1) natural necessity is not directly experienced; (2) observation and experiment yield nothing but contingent descriptions of particulars; (3) from contingent premises no conclusion concerning necessity can be derived; (4) mathematical necessity does not carry over to descriptions of the world. Hume then concludes that necessary connections, even if they do exist in nature, cannot be known to science.[1]

If necessary connections cannot be known to science, then what is left? According to Hume, nothing but particular descriptions and general expectations based on habit. Modern Humeans are not quite so sceptical. They are willing to allow science both particular facts and factual generalizations but, like Hume, they hold that necessary connections between facts, assertions of natural necessity, cannot in principle form any part of scientific knowledge. Therefore, any generalizations which can be established by science, including statements of natural law, must be nothing more than contingent descriptions of the world. That is, if "All *F*'s are *G*'s" is a statement of natural law then, on this view, it can only be taken as the claim that *F*'s are in fact *G*'s, but not as claiming that *F*'s are necessarily *G*'s.[2]

But if scientific knowledge is limited to contingent descriptions, then some recent work suggests that statements of natural law will be unable to serve at least one important function commonly assigned to them in science. Although the matter is too complex to be fully discussed here, it is generally agreed that when "All *F*'s are *G*'s" is taken as a statement of natural law, then we are entitled (among other things) to use the statement as the basis for hypothetical inferences of the form, "If this were an *F*, then it would be a *G*." Indeed, if we were not so entitled, then it would be hard to see what could be meant by calling "All *F*'s are *G*'s" a statement of law. However, it also appears that if "All *F*'s are *G*'s" is only a contingent description it will be unable to support hypothetical inferences, and hence unable to function as laws are supposed to function in scientific practice.

Take "All acids are corrosive" as an example of a (low-level) statement of universal law. On the Humean view the statement can be known by science (on the basis of observation and experiment) only insofar as the statement is understood as a general description of acids, that they are actually in fact corrosive. As Hume put it, science can only show that properties are "conjoined, but never connected."[3] Thus, science can only know that whatever is actually an acid is in point of fact also corrosive. This does provide a basis for prediction since, given the fact that some liquid is actually an acid, it follows that the liquid is (will be found to be) corrosive.

But the law, taken in this way, seems to provide no basis for hypothetical inferences, ones concerning things which are not (or not known to be) acids. Consider, for example, a beaker which is filled with water. Given the law that all acids are corrosive we do not hesitate to conclude hypothetically that "If the water in this beaker were an acid, it would be corrosive." However, it is hard to see how this conclusion would follow logically if the law says only that *actual* acids are corrosive, since the water in question is *not* actually an acid.[4] On the other hand, if the law does apply (as it seems to) not only to things which are actually acids but also to things (such as beakers of water) which are not but might have been acids, then it must say more than the Humeans have supposed. It must say something like "All possible acids are corrosive" or, in other words, "Acids are necessarily corrosive." And in general when a statement of the form "All *F*'s are *G*'s" is taken as a law it seems that we should interpret it as affirming that "*F*'s are necessarily *G*'s".[5]

If statements of law do assert necessary connections and if such statements can be established by science, then science must somehow be able to overcome the Humean limitations to contingent descriptions. The question is how, and that is the topic of the remainder of this paper. I shall begin by briefly discussing two previous attempts to show how science is able to overcome this limitation — one by William Kneale and another by Nicholas Rescher. And I shall end by arguing that Rescher and Kneale, taken together, point the way towards understanding how statements of the form "Necessarily, all *F*'s are *G*'s" can be confirmed on the basis of observation and experiment, contrary to the third point of Hume's argument outlined above.

INDUCTION AND IMPUTATION

In *Probability and Induction* William Kneale concludes a lengthy discussion of the nature of laws by saying that "there is no good reason for denying that laws of nature are principles of necessitation" and indeed that "this is the only account of law ... which appears to make sense."[6] Kneale is, therefore, firmly committed to a necessitarian account of laws.

But then how are statements of law established in science? In a later section of the same book Kneale points out that claiming necessity for a statement of the form "All F's are G's" is equivalent to claiming that it is impossible for an F to fail to be a G. Therefore, he says, "the conjecturing of a law is ... an attempt to say where one of the boundaries of possibility lies." But the fact that something is possible can be definitely shown by finding an instance of it (an F that is not a G), and so a proposed law "may be decisively refuted at any time." Failure to find a refuting instance does not decisively establish a law, however, and in fact Kneale says that a law "cannot be conclusively verified by any accumulation of facts."[7]

Kneale concludes, then, that the acceptance of a proposed law is an act of policy, the "policy of primary induction":

> ... the policy which we follow in the induction of laws consists of two articles: (a) to search for new conjunctions of characters, and (b) to assume the impossibility of conjunctions which are not discovered by continued search.... Our assumption of boundary principles is then an act of policy rather than belief in any ordinary sense, and it is accompanied *at all stages* by a search for evidence which would compel us to revise our hypothesis.[8]

Suppose, for example, that the hypothesis is "Necessarily, all pieces of copper conduct electricity." According to the policy of primary induction we are to search for "new conjunctions of characters," which I take to mean searching for refuting instances of the hypothesis. In this case, then, we are to search for pieces of copper which do not conduct electricity. But where and how and how long are we to search? At what point are we to take the step of assuming the law? Neither the policy nor anything Kneale says later supplies a clear answer to these questions. Evidently the policy needs to be supplemented with some guidelines concerning the nature and duration of a test of lawfulness, a point to which I will return in the last section.

Putting these problems aside for the moment, let us ask how the policy of primary induction can be justified. It is justified, according to Kneale, because "it is the only way of trying to do what we want to do, namely, make true predictions."[9] But even if we extend this to hypothetical inferences, the justification seems rather weak. Hume was perfectly willing to grant that we *want* to make true predictions, only he thought that there was no rationally justifiable way we could do so. In order to justify such a procedure we need to show that it holds out some prospect of success in getting us what we want, that it gives us a good reason for assuming a law as a basis for predictions and hypothetical inferences. But to do this we will need to develop some more specific guidelines for applying the procedure.

Meanwhile, let us consider Rescher's account of the way in which statements of law are established. Rescher agrees with Kneale that statements of universal law are principles of necessitation, and that such statements can be shown to be false by counterinstances. However, Rescher believes that observational and experimental evidence has only a very limited role to play in the establishment of a law. He says that the credibility of the generalization which appears in a law can be raised by empirical evidence, but the same is not true of the additional claim of necessity made in a law.[10] That is, a series of positive instances can help to confirm "All F's are G's," but the same evidence, according to Rescher, can say nothing whatever about whether "Necessarily, all F's are G's." The prefix "Necessarily"—and with it lawful status—is granted to a generalization on the basis of what Rescher calls "transfactual imputation": a commitment by the community of users to treat the generalization as a law and thus endow it with necessity.[11]

This does not mean that laws can be created arbitrarily by an act of individual or collective will. Rescher is very careful to say that the imputation of lawfulness to a generalization must be based in part on a rational warrant:

> The imputation is, to be sure, an overt step for which a decision is required. But to be *justified* this decision must be based on a rational warrant, and must have its grounding in (1) the *empirical evidence* for the generalization at issue in the law, and (2) the *theoretical context* of the generalization.[12]

The rational warrant, as Rescher sees it, consists of observational and experimental evidence for the generalization combined with the ability of the generalization to cohere with other statements in a theoretical structure (a scientific axiom system). But all of this is only a necessary condition for lawfulness, it is not sufficient. A generalization that passes these tests is not thereby a law, but rather just a candidate for lawful status. The granting of that status is a matter of decision or commitment on the part of the community, and Rescher suggests that pragmatic factors such as the usefulness of the statement to science will weigh heavily in that decision. In any case, Rescher maintains, imputation is required since the empirical evidence is always "grossly insufficient to the claim made when a generalization is classed as a law"—the claim of necessity.[13]

In a way Rescher can be seen as filling in some of the gaps in Kneale's account, particularly step (b) in primary induction. As mentioned above, Kneale is not very clear about when we should move on to assuming the law—how and how much of the searching step (a) we should go through first. Rescher is a bit clearer about this. According to him we should go through enough of a search to establish the likelihood of the generalization contained in the law and we should also check its coherence with other scientific generalizations. Once this is done we *may* impute lawfulness to the statement, but we are not compelled to do so. On this point also he is somewhat clearer than Kneale. To assume the statement as a law we must decide not only that its warrant is good enough, but also that there is some practical point (scientific usefulness) in treating the statement as a law.

The main problem that I see with Rescher's account is that in his view the empirical evidence provides no confirmation at all for the claim of necessity. If the evidence were able to suggest, or even hint at, the existence of a necessary connection, then I think that imputing necessity to a statement would be justified. But to impute necessity to a statement without even an evidential hint in that direction is once again to court the Humean objection that we are establishing laws because we want them or need them, not because we know or have good reason to believe they are such. Appealing to theoretical coherence also does not seem to help much on this point since the other statements already in a scientific system have presumably entered the system by way of imputation themselves, and so the problem of justification would be just pushed back a step. On the other hand imputation *is* going to be necessary since in any case we cannot expect the evidence to be logically compelling, but it would be better if the gap between evidence and law were not as great as it appears in Rescher's account. (Rescher is also right about the fact that in scientific practice only statements that have some definite utility are going to be explicitly recognized as laws and given titles such as "The Boyle-Charles Law" or "The Law of Mass Action").

I think that if Kneale's primary induction is supplemented with some guidelines it can be made to provide reasonable basis for imputations of necessity to statements. And if it can I think that the Humean objections mentioned above can also be overcome. I shall try to sketch out such an account in the next section.

Experimenting with Necessity

Let us begin by reconsidering the nature of the problem. We are supposing that "Necessarily, all F's are G's" or, more formally, "$\Box(x)(Fx \supset Gx)$" is a correct model for universal laws of nature. This model is assumed to be equivalent to "It is not possible for an F to be a Non-G" or "$-\Diamond(\exists x)(Fx \ \& \ -Gx)$." The contradictory of both of these statements, "It is possible for an F to be a non-G" or "$\Diamond(\exists x)(Fx \ \& \ -Gx)$" follows from the assertion that any particular F is not a G (e.g., "$Fa \ \& \ -Ga$"). Finally, given all of this, we are asking whether any evidence which might be developed by science is capable of raising the credibility of the necessitarian universal, and not merely by confirming the contingent generalization contained in it.

Since space is limited here I propose to attack the problem in two stages. First I shall present what I take to be an intuitively adequate answer to this question, illustrating it with some simple examples. Afterwards I shall just briefly sketch out a reply to the Humean objections raised above.

There is a perfectly ordinary sense in which we know what is possible and what is not. I know, for example, that it is possible for me to add a green apple to the basket of red apples on the table before me. Therefore, even if "All apples in basket b are red" is true I cannot accept it as lawful. On the other hand I also know that it is not possible for human beings to fly without mechanical aid nor to stop bullets with their bare skin. Therefore, I regard "No-one can fly without mechanical aid" and "No-one can stop bullets with bare skin" as lawful—and I am perfectly willing to add the prefix "necessarily" to either statement.

How do we know what is possible and what is not? As science develops we can, of course, invoke the aid of already established laws and theories. But this cannot be the whole story, otherwise the process of establishing a law would be either circular or regressive. So how can we know what is possible and what is not possible without invoking other laws or theories? Primarily, I want to suggest, through an open-ended use of observation and experiment guided by analogies.

For example, how do I know that it is not possible for humans to fly without mechanical aid? Mainly because it has been tried before a number of times and no-one has even come close to succeeding. Some of the trials made have been based on the closest analogy available, the flight of birds. This would include jumping off of heights, flapping one's arms vigorously, etc. None have succeeded and, furthermore, I know of no other way to try it which offers any hope of success. So long as no new and promising trial presents itself I feel justified in assuming that flying without mechanical aid is not possible. Similarly, I know of no case in which a person has stopped a bullet (of normal size and velocity) with bare skin, and I also know of no promising way in which that might be tried. Of course there *could* be a way unknown to me now, so my commitment is open-ended rather than final. And if a way is found to do either of these things (flying through levitation, for example) I'll recant.

In a more sophisticated vein, the hypothesis "$\Box (x)(Fx \supset Gx)$" can be tested in the following way. (a) Consider any analogy which suggests that an instance of F and non-G is likely to be found or produced somehow. (b) Conduct searches and/or experiments in accordance with these likelihoods. (c) If any such research turns up an F that is a non-G, then the hypothesis is refuted as it stands. (d) If all likelihoods of this sort have been tried and no F that is a non-G has been found, then lawfulness may be imputed to the statement for the time being. Of course the question of lawfulness can always be re-opened at a future time, as soon as another promising mode of trial is discovered. It is important to understand, however, that the question cannot be opened on the basis of *bare* possibility: that there could be a way but no-one has thought of it. (If questions could be reopened on this basis then they could never be considered closed, even provisionally.)

Of course the procedure outlined here is still only a rough approximation of the way in which such hypotheses could be tested. For example, I think that we would not want to accept a law without at least some good tests of type (a), since without them the best we can do is a more or less happenstance sample of F's.[14] Then, too, there is the problem of saying just how likely a research proposal has to be in order to be taken seriously under (a), and there is the problem that some serious proposals may turn out not to be physically possible or economically feasible or morally proper, and so on. Still I think that this outline, rough as it is, captures some important points about testing laws.

First of all it seems clear that following such a procedure to a successful conclusion could raise the credibility of a claim of universal necessity. The reasoning runs like this. When "$\Box(x)(Fx \supset Gx)$" is introduced as a hypothesis it logically conflicts with "$\Diamond(\exists x)(Fx \& -Gx)$"—and so only one of these two statements can be true. The process of confirmation outlined above is designed to maximize the opportunity of verifying the latter hypothesis by finding or producing some F that is not a G. Persistent failure to verify the later hypothesis despite maximum opportunity suggests that it is not true. But if it is not true then it follows that the former hypothesis is true. In other words, evidence which suggests that "$\Diamond(\exists x)(Fx \& -Gx)$" is probably not true thereby also suggests that its contradictory "$\Box(x)(Fx \supset Gx)$" probably is true. Of course the conclusion is always provisional and never at a level of certainty, but it does seem to provide a good reason for adopting a universal law on the basis of what we know at a given time.

If this is at all correct, then Hume's objections would seem to fail at the third point: the claim that a necessary conclusion cannot be based on contingent premises. If this means that we cannot correctly argue from a series of instances of F's that are G's to a necessary connection between being an F and being a G, then Hume is certainly right. But if I am right then this is not the form taken by scientific reasoning in favor of statements of natural necessity. The reasoning is not an attempt to demonstrate the necessary conclusion but rather to decide between it and a verifiable contradictory hypothesis on the basis of the best evidence we can muster at a time. Such reasoning issues not in conclusive proof but rather in a reasonable basis for imputations of lawfulness. There is no reason to think that this process is inherently illogical, and hence no reason to think that science is utterly barred from the realm of necessity.

NOTES

1. David Hume, *An Inquiry Concerning Human Understanding*, ed. by Charles W. Hendel (New York: Liberal Arts Press, 1955). See especially Chapter VII, "On the Idea of Necessary Connection." Hume's arguments there are psychological (concerning the alleged origin of the idea) but they can be easily converted to the epistemological arguments summarized here.

2. See, for example, Ernest Nagel, *The Structure of Science* (New York: Harper, Brace, and World, 1961), pp. 53–70. After arguing (on Humean grounds) that statements of universal law cannot be logically necessary, Nagel claims that they should instead be understood as contingent statements of a type he calls "unrestricted universals" which satisfy certain requirements of evidence. Essentially the same position has been defended by Nelson Goodman, Carl Hempel, and a number of others. It should be mentioned that some Humeans want to take the concept of statistical law as primary and interpret universal laws as limiting cases of statistical law (i.e., as statistical laws with probability approaching 1.0). On this see Wesley Salmon, *Statistical Explanation and Statistical Relevance* (Pittsburgh: University of Pittsburgh Press, 1971) and, more recently, Brian Skyrms, *Causal Necessity* (New Haven: Yale University Press, 1980). Statistical laws will not be discussed here.

3. Hume, p. 85.

4. The argument that contingent generalizations cannot function as premises in hypothetical inferences has been made by William Kneale, *Probability and Induction* (Oxford: The Clarendon Press, 1949), pp. 75–78; by Nicholas Rescher, *Conceptual Idealism* (Oxford: Basil Blackwell, 1973), pp. 54–59; and by Fred Dretske, "Laws of Nature," *Philosophy of Science*, vol. 44 (1977), pp. 248–268. Kneale and Rescher conclude

that laws should be seen as principles of natural necessity while Dretske holds that they should instead be taken as relations of properties on the model "*F*-ness is *G*-ness." Dretske's interpretation of laws has recently been criticized as involving a category error. See Herbert Hochberg, "Natural Necessity and Laws of Nature," *Philosophy of Science*, vol. 48 (1981), pp. 386–399.

5. The phrase "*F*'s are necessarily *G*'s" is meant to suggest the necessitarian universal "$\square(x)(FxGx)$" or equivalently "$-\Diamond(\exists x)(Fx \ \& \ -Gx)$." These formulae could be further interpreted as holding that "All *F*'s are *G*'s" is true in all (physically) possible worlds, roughly in the manner of Nicholas Rescher, *A Theory of Possibility* (Oxford: Basil Blackwell, 1975). However, I shall bypass further questions of interpretation here in order to concentrate on the epistemological problem of the establishment of laws.

6. Kneale, *Probability and Induction*, p. 89.

7. Kneale, p. 227.

8. Kneale, pp. 227–228.

9. Kneale, p. 235.

10. Rescher, *Conceptual Idealism*, pp. 81–82.

11. Rescher, pp. 82–83.

12. Rescher, p. 86.

13. Rescher, p. 82.

14. For much the same reason I think that experiment is generally more useful than observation in corroborating laws. In observation we have to pretty much take what is given — we can only observe the *F*'s that happen to be there at the time the observation is made. If *F*'s of a significant type happen to be missing at that time, so much the worse. But in experiment we can (when successful) produce *F*'s of whatever type seem most significant for testing the law. Experiment, therefore, offers greater control over the testing process.

C.
SCOPE

THE SCOPE OF SCIENCE

Henry E. Kyburg, Jr.

It has been amply demonstrated over the past few hundred years that science provides immensely powerful tools for altering the natural world. Most people would agree that it has also provided understanding and explanation. But it is not at all clear that the tools and methodology of science can lead to resolutions of all of our cognitive quandries. It has been alleged, for example, that scientific knowledge and religious knowledge are so different in character that they cannot even come into conflict. But the recent debates about evolution and creationism suggest that perhaps they can come into conflict after all. It has been argued that the rift between matters of fact and matters of value is so deep that scientific knowledge of matters of fact can have no bearing on our knowledge, if knowledge it be, of matters of value. But it has also been claimed that "ought" implies "can"; and if science yields "can't," that should cancel the obligation of the "ought." Closer to home, there are two issues underlying science itself which are alleged not to admit of any sort of scientific resolution. One is the issue of realism and instrumentalism: Is there any conceivable way in which the tools of science can be brought to bear on this ontological question? The other is the question of the presuppositions or postulates of scientific method. It has been argued that we require substantive empirical assumptions, postulates, or presuppositions — ultimate presuppositions — *before* we can claim soundness for our scientific arguments. If this is so, these presuppositions themselves cannot be defended by scientific argument. The very foundations of science itself may lie beyond the limits of scientific inquiry.

Even if some of these claims concerning the limitations of science are true, they are, as just stated, both cryptic and vague. In what follows I shall expand on them a bit, and I shall argue that in most respects they are defeatist, or self-serving, or arbitrary, or some combination thereof. One way of doing this, of course, would be to argue that all cognitive activity was, *per definitionem*, "scientific." It would follow, all too easily, that insofar as the questions at issue were cognitive, they would admit of "scientific" treatment. Therefore I shall begin by offering a rough and ready characterization of what I take to be "scientific," and the "natural" world to which we apply science. Rather than finding this characterization so loose as to be question-begging, you will probably find it a shockingly narrow, strait-laced, and old fashioned empiricism. Despite the obituaries that have appeared in some of our leading journals, I remain to be convinced that empiricism is dead.

By the natural world, I mean first of all the objects and events of ordinary experience: sticks and stones, trees and flowers, birds and beasts; the succession of day and night, of the seasons; growth and decay, the ordinary behavior of animals and people. Ordinary experience is somewhat vague; the ordinary experience of the eskimo is not that of the south sea islander; and the ordinary experience of the astrophysicist is different yet. But they still have much in common. I take this ordinary experience to be crystallized in ordinary language and its dialects, and in common judgments, and these common judgments to lie at the foundations of science.

There are two ingredients in the development of scientific knowledge. Both have their roots in common sense and ordinary responses to the world. One ingredient is induction, by which I mean statistical inference. If the frequency of A's among a large number of B's has been observed to be f, infer (other things being equal) with an appropriate degree of confidence that the frequency of A's among all B's is approximately f. This form of inference has yet to receive its full articulation—in particular it is not easy to specify what "other things being equal" means, nor how to determine "an appropriate degree of confidence"—but serious efforts to lay bare the logic of such inferences only began a few decades ago.

The other ingredient consists in the reasoned replacement of one way of talking by another. At an elementary, pre-scientific level, an example of this is the development of linguistic apparatus which allows for the articulation of the difference between appearance and reality. At the other end of the spectrum it consists in the introduction of language for talking about the quaint objects and properties important to particle physics. I say "reasoned" replacement; it can be argued that such replacements are made on rational grounds.

II

My belief that both ingredients of the development of scientific knowledge can be represented as rational and articulate processes is an intuition, an article of faith, a speculation. We have neither discovered a satisfactory logic of statistical inference, nor explicit criteria for the replacement of one way of talking by another. This has led a number of writers to conclude that the cogency of scientific argument rests on presuppositions or assumptions. If scientific method itself rests on ultimate presuppositions or assumptions, then those presuppositions themselves cannot be defended by scientific argument: they lie beyond the limits of scientific competence. If the very foundations of science lie beyond the limits of scientific competence, there follow consequences which may or may not be welcome. It is possible to claim, for example, that science and religion can come into no conflict, since they represent alternative points of view resting on alternative presuppositions and assumptions. It is also possible to dismiss scientific knowledge that you would rather not accept by just saying that in that particular regard you decline to accept the presuppositions required.

One of the most commonly alleged "presuppositions" of science concerns causality. This will serve to illustrate my claims. It is said that some assumption regarding causality is required for any scientific inference to be sound, and that (therefore) scientific inquiry is powerless to establish a principle of causality. I contend (1) that no causal principle is needed to establish the soundness of scientific inference, (2) that no principle that has been offered is sufficient to contribute to the soundness of scientific inference, (3) that no principle of causality in a general form is implied by our scientific knowledge, but (4) that if there is such a principle, it can be discovered—rendered acceptable— through empirical scientific inquiry.

With respect to (1): Statistical inference is one standard form of scientific argument. Some writers have argued that some form of causal principle is required for the soundness of statistical inference, but their arguments have been unpersuasive. Given the sketchiness of our present understanding of the logical structure of statistical inference, it is hardly surprising that these arguments are unpersuasive. On the other hand, there are many areas of science in which the major form of scientific knowledge is knowledge of statistical regularity. Whether or not there are assumed to be underlying causal mechanisms, these regularities themselves represent scientific knowledge about the world.

With respect to (2): The claim that there is a causal regularity underlying every event, if stated in barefaced generality, is so loose as to be powerless to aid in the reconstruction of particular inferences. Without some constraints on the sorts of things that can enter into these regularities (and on the sorts of things that can be assumed to be regulated) we can get nowhere. But as soon as the principle is stated in a more concrete and helpful way, for a particular case (e.g., There must be a bacterial or viral organism causing the collection of symptoms we call disease D), it becomes (a) falsifiable, (b) grounded on scientific background knowledge, and (c) quite possibly false (D may be psychological or genetic).

With respect to (3): It seems to me that we have far more evidence for tychism than for the causal principle—things never turn out exactly as we expect them to. With all our scientific knowledge, we can predict with reasonable precision only in special and usually artificial circumstances; and "reasonable precision" already implies that we cannot predict without error under any circumstances. The faith that if only we had more scientific knowledge, we could in principle predict the course of events (at least, the course of non-human events) strikes me as just as childish and ill founded as the somewhat older faith that the course of human destiny is fatalistically determined by the whims of divine beings. It is a useless superstition.

With respect to (4): However, it cannot be denied that science and technology have given us the power to interfere in the natural course of events, to make them conform more nearly to our desires. We have uncovered specific causal connections: bacterium B causes disease D, drug K causes the demise of bacterium B. Leaving to one side the essentially indeterministic character of quantum mechanics, we can *imagine* that there are no bounds on our abilities to discover causal connections. Thus it might be the case that in the limit of scientific inquiry some form of causal principle might be seen to hold true in the world. If that is so, it is something that we can discover only by pursuing scientific inquiry towards its (temporal) limit.

I regard the demand for postulates on which to found science as a failure of nerve. To be sure, we have yet to produce an adequate analysis of either statistical inference, or of the replacement of one theory or law by another. To be sure, it is a speculative hypothesis, or an article of faith, that we will be able to. But if we don't try to produce a presupposition and assumption free analysis of scientific argument, we surely will not find one.

III

There is a lot of discussion in the philosophy of science of the issue of realism versus instrumentalism. Are the entities and processes referred to in our fancier scientific theories *real*—are they out there, doing what our scientific theories say they do—or is the whole edifice of science an enormous black box, that allows us to crank out observational results from observational inputs? Do we not have to settle this question, on philosophical grounds, before we can even understand science?

I suggest that this is putting the cart before the horse. In specific cases, in ordinary discussion, we turn to science for answers to our ontological questions. Are there really electrons? An affirmative answer is mandated by the body of scientific knowledge and evidence we now have. Given the evidence we now have, we have every reason in the world to believe that there are electrons. To say this does not, of course, preclude the possibility that an expanded body of scientific knowledge and evidence might warrant a negative answer. But that is true of any scientific hypothesis.

Is there a deeper question—a blanket issue, so to speak? Can we ask, not of each kind of theoretical entity, but of all possible theoretical entities, whether we have grounds for accepting their reality? The question can be asked, just as the question can be asked whether all our experiences may not be hallucinatory. Like other forms of skeptical questioning, however, it seems doubtful that in this general form the question can be rendered intelligible; and it seems certain to me that there is no way in which the question can be rendered *serious*.

To claim that the question of instrumentalism versus realism can only be asked of one particular case at a time, and that it is scientific inquiry itself that must answer it, is not to say that the answer in any particular case is easy or obvious. The relevant scientific knowledge may be incomplete—there may be some reason to suppose that a theory involving X's is correct, but not strong enough reason to preclude the development of an alternative theory involving entities that act *as if* they were X's. But at some point we may read in *Science* that a forty-eight million dollar experiment at CERN has confirmed the existence of X's. And it is not only the lack of scientific knowledge that may render the answer to the question "real or instrumental?" difficult. Reason and analysis may be required to interpret the deliverances of science. I do not go so far as to say that all ontological questions are internal to the special sciences. I do not go far as to say that philosophy and analysis cannot contribute to tentative answers to these questions. But I think our answers must be tentative, and that the evidence on which those answers should be based is precisely *scientific* evidence.

IV

How about values and duties? Are they too to fall in the domain of natural science? What of Hume's "ought" and "is"? I won't go that far. I won't claim that biology or psychology can tell us what to do or what to value. But at the same time it seems clear that psychology and biology can tell us things that are relevant to our choices and actions. In the first place, of course, science can inform us regarding the consequences of our choices and actions. Rather more to the point, however, psychology and biology can give us insight into what we value and desire. It is naive indeed to say, I know what I want and there is an end on't. On the other hand, it would be equally naive to suppose that scientific inquiry can resolve all questions of value, even leaving to one side questions of duty and obligation. My claim is not that. My claim is rather that we cannot *specify* the limits of science in this regard. We cannot say, so far and no farther. Knowledge of the kind of animal we are, knowledge about our reactions and satisfactions, can help to close the gap between what we think we want and what we really want.

But can scientific inquiry tell us what our duties and obligations are? That's what would confound Hume's observation. Of course the pragmatic imperative: if you want to accomplish A, then you ought to do B, presents no problems; indeed it is one of the main functions of scientific knowledge to provide us with the grounds for accepting such imperatives. Categorical imperatives seem to be quite different. Even here, however, if you accept the dictum that "ought" implies "can," scientific knowledge appears relevant. If at the temporal limit of scientific inquiry, one approached a nomological net that totally constrained human action—that is, that eliminated human action in favor of predictable human behavior—there would be no room for obligation or duty, except as (possibly) useful descriptive terms in that net.

But this is a cheap shot. In the first place, we are nowhere near that limit, and I just got through arguing that it is an unwarranted and useless assumption to suppose that science is approaching it. In the second place, even supposing that La Place's demon lies at the temporal limit of scientific inquiry, that doesn't constrain us now: the nomological net that we know of leaves plenty of room for the notions of duty and obligation. In the third place, we were asking if scientific inquiry could contribute to our deontic knowledge; to say that in some hypothetical limit there would be no duties or obligations is no answer to that question. And in the fourth place, the dictum that "ought" implies "can," and its contrapositive, "can't" implies "not-ought," are questionable.

Nevertheless, even if we weaken the dictum to something like "ought to X" implies "by trying you can come closer to doing X than you could otherwise" what we know of biology and psychology is still relevant to what we can take as obligatory. As in the case of value, it is not clear how far science can go in putting constraints on what we can reasonably take to be our duty; which is to say that it *is* clear that we can *not* specify limits to the relevance of scientific knowledge to our conceptions of duty and obligation.

V

The final region which is often alleged to lie beyond the limits of scientific knowledge is the religious. It is often said that religious questions are beyond the competence of science to answer; it is perhaps most often in connection with religion that allegations concerning the limits of science are invoked. There are a number of issues involved here, largely because the claims of religion can cover such a wide variety of areas.

At one extreme, there are head on collisions between religion and science, as, for example, the conflict between creationism and evolutionary theory. It has been argued that although there is a concrete factual issue involved, it cannot be resolved by conventional scientific procedures. The reason is that conventional science has one set of presuppositions and commitments, while creationism has a different set of presuppositions and commitments. Since these presuppositions and commitments lie beyond the limits of science, there is no non-question begging way of adjudicating the claims of creationism and evolution. Thus A may reasonably believe in creationism and B, with the same evidence, may reasonably believe in evolution.

On the view I am putting forward, this won't wash. The reasonableness of accepting scientific conclusions does not depend on presuppositions or assumptions. Indeed, one of the glories of science, and one of the ingredients of its progress, is precisely its propensity to question, to put to the test, and to repudiate unfounded assumptions. If natural science requires no presuppositions, then clearly one does not arrive at a different body of knowledge by making *different* assumptions; one arrives at superstition, rather than science, by adopting presuppositions that are unwarranted.

There are relatively few who would accept the pronouncements of religion such as the one just mentioned in areas that quite clearly fall in the domain of natural science. There are many more who would accept the results of scientific inquiry, as far as it goes, but who would suppose that there are limits beyond which it cannot go. Beyond these limits there are matters both of fact and of value where religious knowledge supervenes. The Supreme matter of fact, of course, concerns the existence of God. The major matters of value concern Sin and Salvation.

The factual hypothesis is both vague and ambiguous. The idea of divinity is rather amorphous, and every sect has its own god. Nevertheless, it seems to me that it is a factual hypothesis, and that we can go beyond Voltaire's cagey claim, "I have no need of that hypothesis." Negative existential claims are not beyond the purview of science. We have outlived the ether, phlogiston, and a number of older chemical virtues. This is not to say that the theories embodying the assumed existence of these things have been *refuted*; I think the historical and philosophical analyses which suggest that out and out refutation does not occur in science are correct. But as the evidence mounts, there comes a point where it is no longer rational to attempt to operate within the framework of a theory of a certain sort. The evidence can become overwhelming. It seems to me that the factual hypothesis in question falls into this category. Since it is vague, perhaps we can only say that it vaguely conflicts with those hypotheses for which we have good evidence. But conflict is conflict, and it seems as rational to maintain a belief in phlogiston, appropriately diluted so as not to come into conflict with the facts of modern chemistry, as to maintain a belief in the existence of a dilute god who is careful not to intervene in the natural world revealed by science.

One can say much the same of sin and salvation; the existence of heaven and hell; judgment and damnation. Not only have we no use for these hypotheses, but vague and ambiguous as they are, they seem to conflict with what little many of us think we know about value and duty. One could still be very Unitarian, and say that religion is still a source and a repository of important human values. That may be as it may be. Nevertheless, to the extent that scientific knowledge bears on these matters—and I have claimed that it does bear on them—the results of rational scientific inquiry must be given precedence over our initial prejudices and speculations whether those stem from organized religious sources or from sheer idiosyncrasy.

VI

To sum up: I have argued—though perhaps it has been more a matter of simple assertion than of argument—that we can put no limits to the scope of scientific knowledge. In particular, I have argued that religion, ethics, and morals are impinged on by scientific knowledge, and that we can set no limits to those impingements. With regard to science itself, I have claimed that science needs no assumptions or presuppositions, and thus that science has no foundations which are beyond the bounds of scientific inquiry. Logic itself may be construed as the outcome of empirical inquiry in the sense that its useful constraints reflect the interaction of nature and human nature. This is not to say that scientific knowledge embodies all answers to all questions now, or ever. It is to say that I can envisage no question to which scientific knowledge is *in principle* irrelevant.

This is quite different from saying that the correct response to any question is to embark on a scientific inquiry. If I am faced with a moral dilemma, I must resolve it as best I can on the basis of what I know about both morals and the relevant facts now, rather than on the basis of what someone might know about morals some day, given unlimited scientific knowledge. It might be that that knowledge would be relevant if I had it; but I do not have it, and I must do the best I can with what I have. By the same token, having granted factuality to the hypothesis of God's existence, I must admit that natural science might someday employ as its best theory one in which this hypothesis is an ingredient. But I have my doubts. I also doubt that phlogiston will be revived. Predictions as to the course of future science aside, I attempt to believe what the empirical evidence rationally warrants. I take this to be the essence of science, and I take what is warranted at a certain time to be the scientific knowledge of that time. Over all time, I therefore see no limits to scientific knowledge.

SCIENTIFIC METAPHORS AS NECESSARY CONCEPTUAL LIMITATIONS OF SCIENCE

Earl R. MacCormac

Our thesis will be that scientists must use metaphors to postulate new hypotheses and that this necessity does lead in the direction of conceptual relativism where objectivity becomes difficult to discern and defend. The use of metaphors in science, however, does not preclude the existence of literal statements about the world. Yet, disappointingly, these literal statements are not very interesting for science since they arise from our commonsense perceptions of the world. Scientists like all other men can have a direct knowledge of the world through perception (this is not unequivocal) and they can express this knowledge in literal statements. Their measurements of instruments do depend upon direct perceptions; such perceptions have to be linked to theories through experimental procedures that not only are theory-laden, but they are also infected by metaphors used to express hypotheses.

To establish our thesis in this brief compass, we shall explore the use of metaphors by Newton, Faraday, and by the formulators of the concept of the quark. Next, we shall deal with the difference between literal and metaphorical statements finding scientific theories generating a mixture of both. Finally, we shall briefly establish the formation of metaphors as a conceptual process.

Ernan McMullin has described in careful detail Newton's struggle to formulate concepts of matter, gravity and activity, and Newton's efforts to interrelate these entities (McMullin, 1978). In the *Principia* Newton was concerned with the mathematical analysis of laws of force, and with the *physics* of the operations of these laws within nature. Outside the *Principia* and for the rest of his life, Newton was concerned as a natural philosopher to discover the philosophical causes of these forces. Caught in a tension between his newly formulated mathematical physics and the intuitions of common sense, McMullin brilliantly shows how Newton resorted to the formulation of a series of conflicting metaphorical explanations to resolve the conflict. In attempting to answer the question of how matter is moved, Newton vacillated between the Cartesian tradition demanding that motion required contact and the neo-Platonic and alchemical traditions involving a variety of "active principles" to explain motion. Newton's metaphors conjectured to resolve this tension, ranged from that of absolute space viewed as extensionally identical with God, to "aetheral spirits" responsible for the *initiation* of motion in refraction, electrical attraction, gravitation, muscular action, and other such phenomena where new motion clearly appears (McMullin, 1978, p. 77). Newton speculated upon the possibility that the activity of bodies might find its cause in the nature of light. Newton considered the possibility that an "electric spirit" "could bring about the emission of light, besides causing small objects to move" (McMullin, 1978, p. 94). Newton also proposed aether as an active principle causing motion but ran into difficulties when his mechanical metaphors describing the action of aether implied that aether was composed of finite material particles. Adamant that matter itself could not be a source of motion, attributing an active principle of motion to the material particles of aether seemed to contradict this conceptual postulate. McMullin summarizes the two different guiding principles underlying Newton's speculations about how matter is moved as: "(P1) Bodies cannot act at a distance upon one another without the aid of an intermediary"; and "(P2) Matter cannot of itself be the source of new motion" (McMullin, 1978, p. 101). McMullin notes "there can be little doubt that the metaphors of spirit and active principle, with their overtone of a mode of presence in space other than that of material occupation, helped to dissolve the doubts he had expressed about action at a distance" (McMullin, 1978, p. 103). Through McMullin's eyes we see Newton engaging in a kind of conceptual problem solving that requires stretching language through the formulation of novel metaphors.

One might be tempted to conclude that Newton's speculations about the nature of matter and activity that lead him to formulate unusual scientific metaphors do not affect the meaning of the very same terms that appear in the *Principia* arguing that the mathematical and physical meanings stood apart from those of natural philosophy. Such a reply tends to divorce the mathematics of a theory from its physical explanation of the world. In such a separation, we find ourselves back with Newton trying to rationalize two sets of concepts, those of mathematics and those of a physical explanation, into a single unified theory. But these two sets of concepts must be harmonized if we are not to find ourselves in the midst of possible overt contradictions as later Newtonians found themselves with the assumption that space was absolute rather than relative.

Instead of attempting to synthesize the concepts employed by a theory with those applicable to the operations of the physical world, one might try to enforce a separation of the two sets of concepts by developing a set of concepts completely free from theory. Scientists could hardly argue for a conceptual scheme completely apart from the world, for by such a move they would cease to be scientists becoming instead pure mathematicians or creators of fictions. And Michael Faraday undertook to enforce such a divorce by inventing new scientific terms applicable to empirical phenomena and free from any theoretical considerations, or so he hoped (Williams, 1965, pp. 259ff.). To describe the new phenomena that he discovered, Faraday called upon scholars to help him with the invention of new words. Faraday had already formed the words "electrode" by combining the Greek words for "electricity" and "a way," and "electrolyte" drawing again upon the Greek, but this time putting together the Greek words for "electricity" and "I dissolve." Employing the words "exode" and "eisode," however, for the electrical poles used in what we now know as electrochemical plating left Faraday dissatisfied. The Greek connotations of "exode" were "to leave" and of "eisode" the sense of "into" conveying the idea of a process rather than that of a direction which Faraday preferred. Faraday wrote to the Rev. William Whewell of Trinity College, Cambridge upon the advice of a friend. Whewell had become well known for coining new scientific terms having helped among others Charles Lyell by inventing the terms "pliocene," "miocene," and "eocene." Whewell suggested the words "anode" for "eisode" and "cathode" for "exode." At first Faraday resisted these suggestions but they grew upon him and he soon admitted them into his experimental usage and also replaced his word "zetode" by "ion" which Whewell had also suggested.

Pearce Williams presents reasons given in Faraday's correspondence for the acceptance of these words as not only the semantical connotations of the Greek words, but also whether they are pleasing to the ear and will be remembered (Williams, 1965, pp. 262 ff.). At the first suggestion of "cathode" and "anode" Faraday wrote back to Whewell:

> All your names I and my friend approve of or nearly all as to sense and expression but I am frightened by their length and sound when compounded. As you will see I have taken *dexiode* and *skaiode* because they agree best with my natural standard East and West. I like Anode and Cathode better as to sound but all to whom I have shewn them have supposed at first that by *Anode* I meant *No way*. (Williams, 1965, p. 265).

By the invention of new words, neologisms, Faraday and Whewell believed that they could create terms that accurately described experimental facts without a commitment to theory. Filled with semantical overtones, however, these words remained metaphoric in their suggestive juxtaposition of unusual referents. The combination of the Greek for "electricity" and "I dissolve," for example, juxtaposes a physical phenomenon with the action of an agent. Neologisms not recognized as metaphors may become the most dangerous metaphors by hiding the conceptual baggage that they bring with them. Faraday's hope of doing science without theory and Whewell's confidence that definitions of terms should conform as closely as possible to "observed facts," attempt to confine scientific conceptualizations to an intimate relationship with the world, dispensing with more abstract theoretical conceptions often expressed mathematically (Whewell, 1847, p. 12). For our purposes, it is not sufficient to reject Faraday's and Whewell's efforts as violating the modern acknowledgement that scientific facts are inevitably theory-laden, but we must observe that even in such attempts to be empirical and objective, one must construct new concepts expressed in metaphors (neologisms). The need for new linguistic hypotheses even at the empirical level constrains scientific conceptualizations.

The development of the concept of a quark in particle physics presents a vivid series of scientific metaphors that challenge the philosopher to face the issues of how to reconcile our perceptual intuitions of the world with the abstract demands of a theory (Newton's problem), and how to draw a line between theory and observation especially when one admits that all observations are theory-laden (the reverse problem of that faced by Faraday and Whewell).

Quarks are fundamental particles postulated by physicists to bolster atomic theory, particles widely believed to exist and yet thus far unobserved (Glashow, 1975). Quarks through "strong force" interactions compose a set of larger particles called hadrons that include the proton and neutron. Another set of particles called leptons are believed to be *elementary* or *fundamental*. Leptons, however, arise from "weak force" and "electromagnetic" interactions. In offering a theory of how quarks operate, scientists have given metaphoric attributes to quarks including those of "color," "flavor," "strangeness," and "charm." Since quarks could not be literally seen even if they could be isolated, the attribution of color to them occurs in a metaphorical sense. "Flavor" as applied to quarks referred to the types of quark, there being three "flavors" in the original theory given the labels u, d, and s for "up," "down," and "sideways" (Glashow, 1975). "Strangeness" referred to the anomalously long lifetime of the third quark-10^{-10} to 10^{-7} seconds (Schwitters, 1977). Later a fourth flavor of quark was added, that of "charm" with "charm" indicating theoretical properties not found in the other three flavors of quark. Finally, in 1979 evidence was offered for the existence of a fifth flavor of quark (Robinson, 1979). Combinations of quarks and anti-quarks generate the proper quantum numbers for the hadrons. The discovery of the J or psi particle lends some experimental confirmation to the theory of quarks since the psi particle is a meson (hadron) believed to be a combination of a c quark and an anti-c quark and called by some "charmonium" since it possesses net "charm" of zero. Because of the postulated nature of quarks and their existence in combinations that form hadrons, it remains doubtful that quarks will be observed even in the sense of "indirect" observation in the way that other high energy particles have been verified (Shrader-Frechette, 1982). Scientists do claim to have "observed" quarks in the sense of retroduction but admit that quarks have not been made manifest. Shrader-Frechette argues that the claim to have "discovered" charmed quarks depends "upon how one defines 'observation' " (Shrader-Frechette, 1982, p. 140). Unlike Faraday who invented terms to describe experimental phenomena, those who invented the theory of quarks and invested them with attributes did so to unify a theory, and even the terms for which "observation" is claimed remain theory-laden if not theory-overloaded.

One might be tempted to argue that the use of terms like color and charm is perfectly harmless since these are arbitrary labels invoked merely to lend convenience in speaking. Quarks might have been given prime numbers; calling them colors and flavors helps *us* to use them as mnemonic devices. This seems to have been one of Faraday's and Whewell's concerns in picking words that were not too long and were pleasant-sounding for they would be remembered with joy. Different "colors" were given to quarks that were identical in every property except one, namely that property represented by a different "color." This seems analogous to the role that colors often play in commonsense perceptual experience. Consider three chairs identical in every respect except that of color, are they *fundamentally* different or exemplifications of the same piece of furniture? Is color important to their being "chairs" and not "tables"? But is the "color" of a quark a significant property? No clear answer can be given to this question except to note that a theory of massless gluons that carry the strong force and account for changes in the color of quarks has been developed.

The choice of "color," "flavor," "strangeness," and "charm," however, as attributes for quarks does place a conceptual limitation upon the theory. All of these metaphors drawn from ordinary intuitive experience lend an aura of definiteness and correctness to the theory. When we talk about something "colored" even if we know that this is an arbitrary label, we cannot help associating it with something like an object. In this case, a colored quark becomes through metaphoric conceptual juxtaposition like a solid corpuscular object. Such talk remains consistent with the various versions of atomic theory that have reigned for so long in which small units of matter are composed of even smaller units of matter subdivided until one reaches the most fundamental and elementary units, in this case the quarks and the leptons. Shrader-Frechette finds enough anomalies in the quark theory to wonder if high-energy physics has not moved into that stage of science called "extraordinary science" by Kuhn and whether the atomistic account of matter might be overthrown by another account like that of field-theory (Shrader-Frechette, 1977, 1979; see objection by Hendrick and Murphy, 1981). The choice of metaphors employed in describing the attributes of quarks does place conceptual limitations upon the nature of particle physics; the labels we attach to quarks directly affects how we think about them.

These examples of metaphoric usage in science raise the more basic question of whether scientists can ever possess a direct knowledge about the nature of the physical world. Since they must resort to speculative hypothetical metaphors to construct their theories and relate them to the empirical world, does such a methodological necessity confine them to a postulational relativism? This question is a subquestion of the issue of whether there can be objective, literal, observation statements not directly influenced by the theory which they purport to confirm.

There can be literal statements about the world that find their objectivity in intersubjective, almost universal agreements in perception. When we employ ordinary words in their ordinary dictionary senses to describe objects or situations that are publicly available to be perceived, we are speaking literally. This does not mean that literal sentences are precise, or unambiguous. Certainly what we can talk about in the physical world is limited by our bodily perceptual apparatus. This claim for universal perceptual features expressed in language finds partial corroboration in the work of Berlin and Kay on colors (Berlin and Kay, 1969) and in Rosch's extension of their work claiming not only further experimental confirmation of the evolutionary character of color concepts but also evidence for the existence of more prototypical categories of perception than just color (Rosch, 1975a, 1975b, 1976, 1977, 1978).

Our advocacy of the existence of literal statements by which one can judge metaphorical statements to be metaphorical does not eliminate the need for scientists to adopt a basic metaphorical stance upon which to construct their theories. Scientists necessarily adopt, consciously or unconsciously, what Stephen Pepper called a root metaphor (Pepper, 1943) as their most fundamental theoretical presupposition. Many particle physicists seem to assume today that "The-world-is-mathematical" as their root metaphor. Literally, this cannot be true, but they treat the world as if it were mathematical and create theoretical entities like the quark and then search for experimental evidence (observations) to confirm such theoretical postulations. Or, like Faraday, they assume that "Nature-is-an-open-book-whose-terms-must-be-translated." To call science metaphorical in the sense of presuming a root metaphor as a basic hypothesis about the world does not mean that every scientific statement generated by the theory is also metaphorical.

Scientific theories while resting upon a root metaphor can possess literal statements about the world. These literal statements, however, describe commonsense objects and events and do not directly illumine the nature of entities like "force," "ion," or "charmed quark." The fact that literal observations can be made of meters or cloud chambers should give the scientist confidence that he is dealing with the *real, external* physical world even if the entities that he wishes to confirm or disconfirm are only indirectly and remotely related to these observations. In the debate about theory-laden observation terms, the ultimate relation to literal observations, all be they ambiguous and only indirectly related to theory-laden terms which occupy more of our interest, has been forgotten. Most of the observation terms that we seek to confirm or disconfirm are theory-laden—they are parts of a network of theories (Hesse, 1974, chapter 2) and they are metaphoric. The metaphoric character of these theory-laden terms does place conceptual limitations upon what can be tested in experiments and ultimately upon what can be known. Observations of instruments that offer indirect corroboration keeps the scientific endeavor in the physical world and prevents science from collapsing into complete relativism.

The use of scientific metaphors represents the scientist's cognitive effort to produce imaginative hypotheses through the juxtaposition of concepts not normally associated—Newton's effort to link absolute space with the physical extension of God or his effort to find the source of activity in aether; Faraday's linkage of electro-chemical activity with the action of a person; and the association of quarks with flavor, color, strangeness and charm. The formulation of scientific metaphors attempts to speculate about the nature of the unknown in terms of concepts that are well known. New concepts are formed through this juxtaposition of old and new usages. The "old" constrains the way in which we can conceive of the problem and the "new" challenges us to reconceive the problem. Through this tension of old and new concepts the conceptual dimensions of a scientific problem are defined. Such activity, though often only remotely related to the empirical world as in the quark, does not become mere ivory tower speculation. Newton was masterful and courageous in his efforts to reconcile conceptually his mathematical theory with his personal intuitions about the world. Although he failed, Newton challenged us to consider one of the deepest conceptual mysteries of all—why is it that many of our very abstruse mathematical speculations about the world do actually find corroboration in the world? Only the crudest answer can be offered at this point—in some manner the mental speculations which we make about the world, mathematical and metaphoric, do in some way approximate, with all of the conceptual limitations we have noted above, the nature of the physical realm.

REFERENCES

Berlin, B. and Kay, P. (1969). *Basic Color Terms: Their Universality and Evolution* (Berkeley: University of California Press).

Glashow, S. L. (1975). "Quarks with Color and Flavor," *Scientific American*, vol. 233, pp. 38-50.

Hendrick, R. E. and Murphy, A. (1981). "Atomism and the Illusion of Crisis: The Danger of Applying Kuhnian Categories to *Current* Particle Physics," *Philosophy of Science*, vol. 48, pp. 454-468.

Hesse, M. (1974). *The Structure of Scientific Inference* (Berkeley: University of California Press).

McMullin, E. (1978), *Newton on Matter and Activity* (Notre Dame, Indiana: University of Notre Dame Press).

Pepper, S. C. (1943). *World Hypotheses* (Berkeley: University of California Press).

Robinson, A. L. (1979). "New Evidence for Fifth Quark," *Science*, vol. 205, p. 777.

Rosch, E. (1975). "The Nature of Mental Codes for Color Categories," *Journal of Experimental Psychology: Human Perception and Performance*, vol. 1, pp. 303-322.

Rosch, E. (1975). "Universals and Cultural Specifics in Human Categorization," *Cross Cultural Perspectives on Learning*, R. W. Breslin, S. Bochner, and W. J. Lonner (eds.) (New York: Halsted Press).

Rosch, E. et al (1976). "Basic Objects in Natural Categories," *Cognitive Psychology*, vol. 8, pp. 382-439.

Rosch, E. (1977). "Human Categorization," *Advances in Cross-Cultural Psychology*, ed. by N. Warren (London: Academic Press).

Rosch, E. (1978). "Principles of Categorization," *Cognition and Categorization*, E. Rosch and B. B. Lloyd, (Hillsdale, New Jersey: Lawrence Erlbaum Associates).

Schwitters, R. F. (1977). "Fundamental Particles with Charm," *Scientific American*, vol. 237, pp. 56-70.

Shrader-Frechette, K. S. (1977). "Atomism in Crisis: An Analysis of the Current High Energy Paradigm," *Philosophy of Science*, vol. 44, pp. 409-440.

Shrader-Frechette, K. S. (1979). "High Energy Models and the Ontological Status of the Quark," *Synthese*, vol. 42, pp. 173-189.

Shrader-Frechette, K. S. (1982). "Quark Quantum Numbers and the Problem of Microphysical Observation," *Synthese*, vol. 50, pp. 125-145.

Williams, L. P. (1965). *Michael Faraday* (New York: Basic Books).

Whewell, W. (1847). *The Philosophy of the Inductive Sciences*, Vol. II, (New York: Johnson Reprint, 1967 edition).

ROMANTIC SCIENCE AND THE FUTURE OF THE KNOWING SUBJECT

Joseph L. Esposito

I want to consider whether it makes sense to argue that there are limits to what science can eventually discover in the future if that argument is (1) based on what we now scientifically know, or (2) based upon extra-scientific (in particular, philosophical) theorizing. I deny that the former is possible, but not the latter, and suggest one potential scenario illustrating (2).

Do we now have scientific grounds for claiming there are limits to scientific knowledge? If it is argued that at some time in the future we would be in a position to scientifically ascertain limits to scientific knowledge, then that knowledge would now have to have some scientific plausibility if the claim is not to amount to bald conjecture. For example, if we assume that science is itself an "expression" of certain sorts of relations among material things, then the more we learn about those relations the more, so it seems, we would be able to hypothesize about what kinds of relations are and are not possible in the real world. Just as the study of the physiology of perception gives us a scientific basis for claiming that certain forms of radiation cannot be perceived directly, so the study of the entire "knowing apparatus" of science (including all endo- and exosomatic structures) should be able to give us a scientific basis for claiming that certain aspects of nature also could not become known.

Yet is there a parallel between "limits of perception" and "limits of knowledge"? We now have scientific grounds for claiming there are limits of perception, for we have acquired exosomatic (instrumental) means of scientifically discoursing about all forms of radiation, and we have physical theories that subsume physiology of perception. On the other hand, in physics we still have scientifically respectable prohibitions to ascertaining certain facts dictated by the uncertainty principle, quantum mechanics, and relativity theory. In the latter case, for example, it is currently good science to claim that we can never establish the simultaneity of distant events without the assistance of certain assumptions whose truth cannot be *scientifically* established. Let us say, then, that at a time before Einstein a group of scholars assembled to debate the question of the limits of future scientific knowledge and one of them proposed a restriction on our knowledge of distant simultaneity. At the time such a proposal could not have had a scientific basis. At present we accept the restrictions delineated by relativity theory, but does that tell us what the future course of scientific knowledge will be? Certainly we cannot, with scientific respectability, suggest that because our prescient scholar was previously correct, that we are in a similar position to claim certain limits on our own knowledge and that of future scientists. All we can say is that so far as we currently know we will never be able to know two widely-spaced events occurred simultaneously when at a far distance from us. For to say otherwise is to rule out new sources of knowledge (perhaps telepathy) and to deny the possibility of new languages, categories, timescales whereby the present facts, as we comprehend them, would be able to be described by future scientists not as limitations on knowledge but as rich sources of many new facts.

In other words, claims about the limitations of future science are either scientific hypotheses and so are really about present science, or else they are not scientific claims at all, but either pure fancies or heuristic devices for discovering some new and interesting scientific problems. An example of the latter might be the suggestion that there is an upper limit to the amount of information the human brain is capable of storing, so that no matter how powerful our computers could become, their usefulness would be limited by our ability to handle the complex languages needed to produce and exploit them. This suggestion then leads to the problems of how now to measure brain storage capacity, and whether continuing study will eventually suggest implications about *future* knowledge. To solve these problems we must investigate what we know now, and as our questions lead to new experiments and our experiments to new information we will most likely increase, perhaps vastly, our knowledge of the storage and retrieval capacity of the brain. Yet at no time will the information we acquire enable us to say that an upper limit has been attained for all time.

It would seem then that the question of the limitation of scientific knowledge is not a question that science, as we know it, is in a position to answer. True enough, science is continually aware of its *present* limitations, for without this awareness there would be no drive to hypothesize further; yet it can recognize its limitations only as specific empirical truths continually open to revision by future knowledge which must in principle be regarded as unlimited in scope. This seems to be one reason why the recent interest in the history of science has had no significant impact on scientific methodology or the core conceptions of rationality in the natural sciences. Contemporary scientists may have become more tolerant of rival theories as a result of looking at the consequences of past controversies, but when they construct their experiments, and formulate and defend their hypotheses, these other issues become moot. History of science will become recognized as scientifically significant only when the subject the scientist studies (nature) is itself affected by human *historical* influences (as in the case of the social sciences).

Turning now to (2). If science is and must be historically blind, this is not the case with philosophers and social theorists who look at science from without as an activity embedded in a real social world, dependent upon real individual minds and bodies and the conditions that make them function. In this arena there is wide room for speculation on actual or possible historical conditions that could limit scientific knowledge. Here most prominent consideration in recent years has been given to the socioeconomic conditions of scientific research as an institutional activity.[1] But equally interesting is the investigation of the philosophico-ideological conditions which establish and sustain the identity of the scientist as a human, inquiring person.

Let us trace the development of the scientist. A person comes to think of himself as a scientist after he has learned certain ways of thinking about certain kinds of problems. His education consists of a historical recapitulation of the fruits of the great historical scientific personalities, combined with a study of the work of the great mathematicians. The former provides the content, the latter much of the form of scientific thinking. Then upon entering graduate school the student is ready to begin his own scientific work. What that student has learned during this process, mainly by replaying the original mental dramas of past scientists, is how to exercise his own imagination so that the problem situation under consideration becomes sufficiently idealized to allow for mathematical interpretation. Here Popper would say that the student has merely been arranging the available furniture of the Third World, though the matter is not that simple, for Popper does not tell us how the Third World changes or emerges at any given time.[2] Instead he often seems to rely upon belief in a fixed Platonic World that undergirds it.[3]

Popper's motivation, of course, is to establish the autonomy of the Third World so that we face the imperative of rational criticism in science, and recognize the possibility of unanticipated consequences (i.e., feedback) from that encounter. But he fails to distinguish between the Third World of stored exosomatic information and a purely Imaginative World the practicing scientist retreats to when he is in the process of constructing a hypothesis.[4] True enough, the Imaginative World continuously borrows material from the Third World, but it also remains in touch with the Second World in that its *raison d'etre* is that it can be an expression of a will-to-believe some possible truth about the world on the part of an individual knower. It is not that the scientist wants to find something to believe in — the impression Popper gives in his disparaging remarks about "belief philosophy" — but rather that he wants to find a solution worthy of his assent.

The Imaginative World has two essential characteristics. Its contents must be *signs*, some of which having hypothetical (or referential) significance, and its configuration must be a *constructive arrangement*. All artistic productions belong to this world, and the scientist, while he works in it, does the work of an artist. The provisionally finished result, is an *image* of possible reality the scientist can use as a measure with which to communicate his thoughts or to construct a physical apparatus. Present day functionalists would emphasize the intentional nature of the contents of this world;[5] others would call it the result of sleepwalking, or the work of genius.[6] On the subject of abductive inference, for example, C. S. Peirce observed: "Abduction consists in studying facts and devising a theory to explain them. Its only justification is that if we are ever to understand things at all, it must be in that way."[7] I take this to mean that there could be no deductive justification of abduction. (Peirce occasionally did suggest "justifications" for this process, but to be consistent he would have to admit that these too were abductively generated hypotheses about the conditions of abduction.) It would seem, then, that the Imaginative World has a transcendental character that makes any attempt to subjugate it stultifying.

This emphasis on the imagination will doubtless invite the suggestion that I am proposing a scientific counterpart to the nineteenth-century romantic theory of artistic creation.[8] And to a certain extent this is true. The Imaginative World cannot be explained without what the nineteenth-century romantic philosophers called the "transcendental ego," that is, the self struggling to gain control of its thoughts the way the artist struggles to master his materials. This does not mean that the "material constituents" — so to speak — of the Imaginative World (images, symbolic relations, forms and structures, etc.) are any more the exclusive product of the scientist than are the materials of the artist a product of the latter's own creativity. What is original is how the materials are given their significance and how those signs are combined into a picture or configuration that gives the scientist a sense of achievement or completion.

In seeming contrast to this romantic view, at least two other procedures are conceivable for the formulation of testable hypotheses. One involves dialogue between two or more human minds where the burden of elaborating the Imaginative World is shared by a kind of groupthink of the members. In this case no member has time to think too deeply on the matter at hand or comprehend fully what is being proposed; yet the analysis advances because the possibility of fruitful associations is greatly enhanced by the interaction of the cogitators involved. The other procedure involves the interaction of individual minds with computers having either powerful sign-producing ability or powerful heuristic algorithms.[9]

Popper, I think, would have us believe that all three approaches are really the same: each involves the interaction of a single mind with the autonomous Third World of objective knowledge, so that the Faustian isolation of the romantic scientist is merely an illusion. Actually I think that the other procedures are at least compatible with the romantic method, and, perhaps hopefully, may be newer manifestations of it. For example, for groupthink to be scientifically worthwhile, the give-and-take must lead to new interpretations and images that can be comprehended by each of the participants in terms of their *own* Imaginative World. Otherwise, not only would communication and consensus be illusory, but no critical accounting would have taken place during the dialogue. In the case of computer simulation, the scientist has a powerful associative tool for manufacturing furniture for the Imaginative World, but he must still do the work of an "interior designer" in converting the results provided into a constructive representation that is relevant to *his* problem, something the computer presumably knows nothing about. In both cases, then, the Imaginative World, with its transcendental foundation and maker, cannot be dispensed with.

The romantic method has carried science a long way since Newton first turned that falling apple into the sign of a vector. But, to return now to our question: how indispensible is it to the future course of science? Can scientific rationality be carried along quite adequately by means of a groupthink and computer simulation that is not *continually* tested for relevance against the constructions of the Imaginative World? This question may have only a conditional answer, for, as argued, the science of every age is regarded by contemporaries as being fully 'science' just as we now regard science as worthy of its name even if it does not cure cancer or produce perpetual motion machines. So for purposes of argument I will say that a science *progresses* if previous generations of scientists would have been impressed with its results, and it *regresses* if those same scientists would have judged it worse than their own.

With these qualifications in mind, then, I am claiming that a science without the Imaginative World and its self-conscious, romantic methodology is likely to eventually become a regressive science in the sense just given. Such a science would be marked by a fixed and frozen Third World and by the demise of critical and creative ability. The golden era of romantic science from the seventeenth to twentieth centuries, with its Hellenic ability of reflective thought and imagination, would give way to a Hellenistic science that could only bring to perfection the established techniques of the previous age and then repeat them endlessly and on a more gargantuan scale. This is the kind of situation I imagine would result if we were to let computers do more and more of our scientific thinking for us. This on a large scale is the kind of problem we already confront with the use of diagnostic computer programs in medicine. At its extreme we might imagine a computer producing misdiagnoses of patients and then storing the misunderstood information from those cases for use as symptoms in future misdiagnoses.[10]

Computers themselves, however, are not the greatest threat to the Imaginative World. They might at some point become indispensible if the products of the Imaginative World—the theoretical models of science—become increasingly accessible only to a few experienced practitioners, and "knowledge" takes the form only of computational procedures for the novice; but such a situation is likely to lead to a deceleration of science itself, and then perhaps hopefully to a reemergence of the Imaginative World.[11] As I see it, the problem for us is not merely the demise of the mental practice of theorizing, but also the rise of certain conditions and/or theories that countenance the decline of theorizing. For example, the decline of the Imaginative World follows directly upon the decline of the moral and social status of the critical, creative self. At first sight it seems paradoxical that a self could decline in selfhood and still remain a socially recognizable human being.[12] But this may be quite a simple matter of losing, by cultural degeneration, habits and categories of thought prevalent in previous generations. Today the reflective feats of Plato are possible to any student who takes his or her first course in philosoph, and yet a generation of Platos could not guarantee the continuation of their reflective abilities without zealous support for the mundane conditions and processes of education. Or decline of selfhood might be a more complicated matter of the emergence of complex theories that diminish the standing of the act of knowing by providing a rationale for believing that thinking is always the result of specific conditions in the brain of the knower. The more the human brain becomes looked upon as like a vast computer the more its products may look like just so much data, and the less will the Imaginative World be regarded as a vehicle of discovery.[13]

Against this it might be argued the science as we know it has been so good to us we would never willingly give it up. The ability of the scientific imagination to "deliver the goods," to solve problems people want solved, will be enough to keep the Imaginative World inviolable. Hopefully this is so.[14] Yet even here if our wants also are corruptable by becoming unrealistic and inconsistent, then even romantic science could not save us. If the Imaginative World were to rely upon the actual practice of science for its vitality, I think it could and ultimately would be subject to the vicissitudes of the men and world that science itself serves. Fortunately, the same need not be said about that safe haven for the imagination represented by philosophy, and by epistemology in particular. If we think of epistemology as the act of imaginatively reconstructing important features of ordinary experience with the aim of discovering what is constant and what is variable in it, and if we follow Popper[15] in calling epistemology the theory of scientific knowledge, then the continuation of epistemology guarantees the continuation of an Imaginative World that is at least conducive to scientific thought. On the other hand, if it should be possible to conceive of an epistemology without a knowing subject, then we are likely also to have before long a Hellenistic science without the knowing scientist.

In broad outline, then we seem to be faced with at least four possibilities. Either science will never know (in any significant way) the "knowing" subject, and the Imaginative World will continue on, potently, mysteriously, and inviolable, vulnerable only to social catastrophe; or science will be self-limiting in that the more the knowing subject comes to be known the less the progress that made that knowing possible in the first place can continue; or the procedures of imaginative-world construction somehow will become externalized in such a manner that science will no longer need to rely upon individual persons formulating individual hypotheses as the vehicle of scientific discovery; or, finally, the very value of science-as-usual will come to be questioned on moral, social, and aesthetic grounds and future generations will choose to live in a "post-scientific" age where computers will be instructed to randomly throw away segments of the vast store of knowledge only so that the traditional practices may be maintained, and the need for the Imaginative World artificially or self-consciously kept alive so as to maintain the dignity of the knowing subject.

NOTES

1. Karin D. Knorr-Cetina, *The Manufacture of Knowledge* (Oxford: Pergamon, 1981); Yehuda Elkana, et al eds., *Toward a Metric of Science* (New York: John Wiley, 1978); Nicholas Rescher, *Scientific Progress* (Oxford: Blackwell, 1978).
2. For a criticism of Popper along these lines see Susan Haack, "Epistemology With a Knowing Subject," *The Review of Metaphysics*, vol. 33 (1979), pp. 309-35.
3. Karl Popper, *Objective Knowledge* (Oxford: Clarendon Press, 1972), p. 116.
4. This is not quite, though close to, a bifurcation of the Third World by E. D. Klemke into signs and cognitive contents; see "Karl Popper, Objective Knowledge, and the *Third World*," *Philosophia*, vol. 9 (1979), pp. 45-62.
5. Daniel Dennett, *Brainstorms* (Bradford Books, 1978), pp. 174-189.
6. In *Faces of Science*, ed. by R. G. Colodny (Philadelphia: ISI Press, 1981) V. V. V. Nalimov observes (p. 5): "The first paradox in the development of science is the creative constituent. The process of formulating novel hypotheses does not possess any traits unique to science. In any case, it cannot be distinguished from myth creating."
7. Charles Peirce, *Collected Papers*, ed. by C. Hartshorne and P. Weiss (Cambridge, Mass.: Harvard University Press, 1960), vol. 5, para. 145.
8. Joseph Agassi discusses genius in the context of "romantic science" in *Science and Society* (Dordrecht-Holland: D. Reidel, 1981), pp. 192-209.
9. H. A. Simon et al., "Scientific Discovery as Problem Solving," *Synthese*, vol. 47 (1981), pp. 1-27.
10. On such danger see Kenneth Schaffner, "Modeling Medical Diagnosis: Logical and Computer Approaches," *Synthese*, vol. 47 (1981), pp. 163-99.
11. Rescher, pp. 234-38.
12. See my "Systems, Holons, and Persons," *International Philosophical Quarterly*, vol. 16 (1976), pp. 219-36.
13. This is the double-edged sword of Mario Bunge's analysis of mind in *The Mind-Body Problem* (Oxford: Pergamon Press, 1980), Chs. 7-8. Perhaps the same may be said for functionalism which allows us to keep the old-time descriptions, but then tells us they don't have the descriptive force we previously thought they had.
14. Nicholas Rescher, "Scientific Truth and the Arbitrament of Praxis," *Nous*, vol. 14 (1980), pp. 59-74.
15. Popper, p. 108.
16. See my "Some Grounds for a Moral Criticism of Science," *The Southern Journal of Philosophy*, vol. 13 (1975), pp. 47-54.

D.
LIMITS

FORMS OF ORGANIZATION
AND THE INCOMPLETABILITY
OF SCIENCE

William Bechtel

Within the "unified science" tradition in philosophy of science, the completability of the scientific enterprise was often thought to depend only on the completability of physics (or whatever the lowest level science turned out to be). The possibility of developing a comprehensive and universal physical theory is itself a major issue, but this paper will address itself to another issue—whether, given a comprehensive theory at one level of science,[1] we could even then complete the scientific enterprise. This enterprise I construe rather traditionally as one of explaining and predicting the events in nature. I will argue that even with a comprehensive lower level science (one that provides laws governing entities at one level in nature) one will not be able to explain and predict all events in nature. This thesis would be of little surprise to a non-physicalist, but I will argue for it within the context of a physicalism that assumes all events are physical events or compounded from physical events. The thesis to be argued here must be kept distinct from the more common and weaker thesis that even with a comprehensive lower level science there will remain laws describing the phenomena that will remain unknown. I will discuss this weaker view in the first section and then show in subsequent sections how attention to levels of organization in nature and their effects leads to the stronger conclusion that, as long as higher levels of organization remain unknown and their laws undiscovered, lower level events themselves will remain inexplicable and unpredictable even given knowledge of lower level laws.

Theory Reduction and Special Sciences

It is the "classical" account of theory reduction that gives the greatest support to the idea that science itself will be completable if physics is completable. Reduction, according to this view, requires two conditions — connectibility and derivability. One first had to connect the expressions in the theories of the reduced science with those of the reducing science (this connection was often taken to involve establishing bridge laws that were biconditionals linking the expressions in the two theories). One then had to derive the laws of the reduced science from those of the reducing science (Cf. Nagel, 1961). Many have criticized this account of reduction and proposed others (Cf. Schaffner, 1967 and 1980, and Wimsatt, 1978) but here I will examine only the criticisms Fodor (1974 and 1979) makes in the context of his defense of the autonomy of the special sciences.

Within the "unified science" tradition, reduction was taken to serve two functions — providing ontological simplicity and explanatory unity. If higher level sciences (or what Fodor calls "special sciences") were reducible to lower level sciences in the manner of this classical model, then they would be unified with physics both in terms of their ontological claims and in terms of their explanatory capacity. Fodor proposes an alternative model of the relation of theories that still satisfies the objective of ontological simplicity but denies the explanatory unity claimed by the classical model of reduction. He preserves the ontological claim through what he calls token physicalism. This position claims that every entity that satisfies a predicate of the higher level sciences also satisfies some predicate of the lower level science.[2] Fodor denies, however, that the predicates themselves can be equated. Fodor takes the predicates of a theory to define types and claims that often entities of a type for one theory fall into several different types in a theory at another level. Entities from many lower level types can, for example, belong to the higher level type "money," and other members of the same lower level types may not belong to that same higher level type. Thus, the relations between types at different levels is many-many and not one-one. The strongest inter-level identity one can establish is a token-token identity, not a type-type one. Because the types that are referred to in the laws of the special sciences do not map onto the types specified in the lower level laws, Fodor contends that one cannot equate the predicates in lower and higher level theories. Thus, one cannot derive the higher level laws from lower level ones and so the explanatory power of higher level laws remains anomalous from the perspective of the lower level.

Richardson (1979 and 1982) counters claims like Fodor's against the classical model by showing that the classical model is not itself committed to biconditional bridge laws that generate a type-type mapping between different sciences. All the traditional model does require, he contends, is that, in the lower level theory, there be nomologically sufficient conditions for determining to what type, as defined by the higher level theory, a given entity will belong. There can be a many-many mapping between the types of the lower level theory and those of the higher level theory and still be a reduction as long as such conditions can be specified. In essence, what this requires is that one be able to specify boundary conditions under which a given entity described in terms of the lower level science will be of a type specified in terms of the upper level science.[3] Richardson then contends that the arguments for the autonomy of functional psychology that have been advanced by Fodor and others do not establish any such autonomy. Even given many-many mappings between the types employed in different level sciences one can still explain the laws presented in the higher level science in terms of the lower-level science.

Levels, Causal Networks, and Higher Level Constraints

The discussion of the previous section was couched in the language of theories and theory derivation. Couching such a discussion in terms of theories and the derivation of theories has become traditional in philosophy of science, probably due to the attention paid to sciences like physics and chemistry, where formalized theories are common and where developing derivations from such theories is a principal activity. In biology and psychology, however, there are a number of fields that do not allow, at least at this time, for the development of such comprehensive and abstract theories as one finds in theoretical physics and chemistry. The area in biology best studied by philosophers of science, not surprisingly, is the one area in which such logically and mathematically articulated theories have been proposed— evolutionary theory, especially population genetics (cf. Williams, 1970). But if one looks carefully at other areas in biology, for example physiology, one finds a quite different kind of research. While sometimes appealing to general theories (e.g., those of enzyme kinetics) physiologists and biochemists are more typically trying to build models to explain how certain functions are performed in a living body. Such models specify the objects involved in the system and how they interact to produce the system's behavior. This account is itself taken to be the explanation—it need not be derivable from axioms. The explanatory task for researchers in these areas just is to describe how components of a system interact to produce the effects one seeks to explain. The models they build are meant to describe these interactions.

Scientists in fields like biological chemistry and psychobiology carry out an activity that goes under the name of reduction. The analysis of reduction that focuses on the derivational relation between theories, however, does not help to explain what these scientists are doing since their theories lack the formal structure in terms of which such deductions can be made. What these scientists use reduction for is to explain the behavior of an object that is relevant to one science in terms of objects studied by another science. To understand the activity of reduction as it is pursued by these scientists one has to focus on the way objects are being connected in the models the researchers develop.

What makes this activity reduction as opposed to other forms of causal explanation is the fact that the objects appealed to in explaining the behavior of one object are usually parts of that object. Thus, one is dealing with a part-whole relation and trying to account for the behavior of a whole in terms of the behavior of its parts. In the context of a part-whole relation, it becomes natural to speak of the parts as lower level entities compared to the object they jointly comprise. Recognizing the ubiquity of the part-whole relation in nature, one is led to conceive of nature as consisting of a nested hierarchy of part-whole relations. It is conceivable that in each context the relations of parts to whole might be different so that one could not define levels across nature. Wimsatt (1976), however, has suggested that there are levels of entities across the natural world. In his portrayal, a level is defined by a set of objects that have their primary or most important causal interactions with each other. Typically, an object will only interact causally with an object at another level indirectly, either as its parts interact with that object or as the whole of which it is a part interacts with the object.[4]

This notion of a hierarchy of levels fits naturally with the hierarchy of sciences — scientific disciplines seek to explain and predict the behavior and causal interaction of objects at one of these levels in nature. By tying the hierarchy of sciences to a hierarchy of levels in nature, we have a natural way to understand the role of reduction. When one wants to understand how an entity is able to perform as it does, one has to inquire how it is put together from parts and how these parts contribute to the behavior of the object. One assumes that it will be several entities at the lower level that will constitute the upper level entity and, through their interaction, produce the behavior of that entity. The task is to discover empirically what are the parts whose interaction produces the object's behavior and to describe this interaction itself.

Within this perspective on reduction it is clear immediately that bridge laws or one-one mappings between types of entities is unnecessary.[5] The case is much closer to that described by Richardson where all one needs is to have nomologically sufficient conditions at the lower level that determine the type of higher level entity that is produced. In part-whole terms, one needs nomological conditions fixing what kind of whole the parts will constitute. There is nothing requiring the same set of objects always to form the same higher level entity. If they interact differently between themselves or with other objects, the same parts may form different higher level objects that behave differently. As well, there could be wholes that behave the same but that are comprised of different types of parts.

So far this alternative construal of reduction seems quite compatible with the possibility that discovering all the principles governing the lower level entities will generate a complete science of nature. We may even assume that knowing all the principles governing the parts will tell us how the wholes will behave. But that assumption is wrong. In discussing Richardson's defense of the classical view of reduction earlier I glossed his view by saying that we avoided any problem many-many mappings between levels posed for reduction by knowing the boundary conditions that determined what upper level effect a lower entity produced. The role of boundary conditions, however, is absolutely pivotal and the label "boundary conditions" may mask their significance. The reason a lower level entity can have multiple upper level effects is that its behavior differs depending on the causal network in which it stands. The reason two lower level objects can produce the same higher level effect is that they can respond similarly to the particular causal connections realized in the network in which they stand. What we often call "boundary conditions" from the perspective of a theoretician working out mathematical relations will appear as networks of causal interactions from the perspective of someone attending to the behavior of actual objects. I contend that it is these organized causal networks that play the crucial role in connecting lower level and higher level objects. The effect a given entity has on another entity is partly determined by the way the first entity is embedded in a causal network. (E.g., a carbon atom bound in glucose has different causal efficacy than one in carbon dioxide.)

So far I have only claimed that the causal networks in which objects appear may affect the upper level effects an object may have—the network is important in determining the behavior of the whole of which something is a part. But the potentially more important consequence of the part being embedded in a causal network is that the network may affect the part's own behavior. The way an entity behaves depends not only on the entity's nature but on the causal inputs to the entity. Change those inputs, and the entity will behave differently. The causal network in which the part is embedded is providing inputs to the part and so is capable of regulating its behavior. It was largely to capture this point that Bernard (1865) introduced the concept of an "internal environment" into physiology. Vitalists had long appealed to the irregular behavior of living organisms to argue against mechanists in physiology. Bernard's answer was to point out that the behavior of any organ in the body depended on the inputs that organ receives from other organs in the body (its internal environment) and so any change in these other organs could change the way the one organ and thus the whole organism responded to an external stimulus.[6] So one must know the causal network at the higher level to explain and predict behavior of objects even at the lower level.

One might object, at this point, that the effect of a causal network on a component embedded in that network is no different than any other causal input and so will be handled in the lower level science in the same manner as these other causal inputs. This, however, is a serious methodological error. The effects of an organized system on its members is qualitatively different from that of stimuli outside the system. Since Bernard, physiologists have come to realize just how self-regulating physiological systems are. For example, metabolic pathways are built of allosteric enzymes that modulate the pathway's behavior. The regulating feedback processes within a self-regulating system have a pattern to them that external stimuli typically lack. The feedback stimuli are partly a product of the component's own activity. Thus, such feedback inputs are appropriately viewed as intrinsic aspects of a component's operation, not just as inputs from outside.

ORGANIZATIONAL NETWORKS AND THE INCOMPLETABILITY OF SCIENCE

Once one accepts the claim that the causal networks in which an object is embedded can regulate both the behavior of the object and the higher level effects produced by that object, one needs to address the question of how these causal networks are to be studied. If there were no way to identify these causal networks, all one could do would be to try to average over their effects and so to try to develop one's account of nature at the lower level. Insofar as there are local differences between the networks in which an object can be embedded, though, one's explanations and predictions would be seriously incomplete or inadequate. If, on the other hand, we can identify and map the causal networks at the higher levels, we can hope to account for these local differences and explain and predict local phenomena.[7]

If there are local differences in the causal network as viewed from the lower level, these might be due to objects identifiable at the higher level — entities with sufficient coherence and integrity that we can distinguish them from the ongoing background events. Whether there are such objects identifiable at an upper level depends on whether nature is organized in accord with what Simon (1969) calls "near-decomposability." In a nearly decomposable system, the whole is built out of subsystems whose parts causally interact principally with each other and only to a much lesser extent with the parts of other subsystems. Simon uses the idea of subsystems that are semi-stable to show how one can resolve an otherwise serious design problem — without such subsystems a designer or evolution would have to put the whole system together at once. With decomposability, though, parts can be engineered or evolve independently and subsequently be combined.[8]

Near decomposability not only makes evolution or construction of larger systems possible, it also makes explanation possible. If one can analyze a whole into sub-systems, one can divide the explanatory task. One task is to discover the interaction of the subsystems; the other is to explain each subsystem's individual behavior. These tasks can be pursued with some degree of autonomy from each other. One inquiry can empirically study the interaction of the subsystems while the other can try to determine their capacities for interaction by inquirying into their internal organization. While the autonomy between these inquiries is not complete (and should not be treated as strongly as some advocates of autonomous psychology have advocated[9]), particular researchers can focus their energy on one or the other of these tasks. These sub-tasks are much more tractable than the whole task would be if there were no higher level subsystems and all explanation had to appeal to the interactions of basic atoms with one another under each set of circumstances in order to explain the phenomena of nature. We have no guarantee, of course, that nature consists of higher level objects that form subsystems of the whole in such a way as to facilitate explanation, but the success of dividing the explanatory work of science between different disciplines, each focusing on a different level of organization, suggests such objects do exist (cf. Wimsatt, 1976).

One aspect of Simon's account of evolution through stable sub-assemblies is misleading. It suggests that parts are first crafted for a task and then connected. But it is well established that nature typically replicates an existing type of part and then the redundant parts become specialized to fulfill different needs that exist in a system. Thus, the pattern is that the system constrains what happens to the parts, shaping them to perform specific functions. Sometimes the effect of this is to make the parts interdependent and not self-sufficient. This destroys the potential for analytically decomposing the system. I shall discuss a consequence of this for the division of labor between scientific disciplines below.

Even given the qualification of the previous paragraph, phenomena in nature do seem to be, to a significant degree, decomposable so as to allow a division of labor between higher and lower level sciences. In accord with this division of labor, it is part of the task of higher level or special sciences to identify the causal networks involving higher level objects as they occur in nature and to discover their effects. If these networks are themselves stable, then we can identify yet higher level objects constituted by these networks. Thus, Bernard treated a living body as such a higher level object and argued that physiology has the task of discovering the organizational properties that made the body definable as a stable object. The body, of course, is made of chemicals and *in vitro* experiments could show the possible reactions the chemicals of the body can perform. But, for Bernard, physiology had to determine how the internal environment was organized. This organization he took to be important since it fixed the conditions for each reaction in the body. Since chemical reactions are sensitive to these conditions, the organization partly determined which reactions would actually occur. Discovering the organization within the body became a central theme of physiology. When biochemistry began to develop after Buchner achieved cell-free fermentation in 1897, many chemists thought that the study of individual reactions could answer all the questions of metabolism. Hopkins (1938), however, continually stressed the importance of studying the organization within living systems and his appeals were finally vindicated in the 1930s when, with the discovery of cyclic pathways, biochemists came to appreciate that chemical reactions always occur in the body in a highly structured and self-regulating environment (cf. Fruton, 1972 and Bechtel, forthcoming *d*).[10]

I contend that it follows from this conception of the way in which organizational networks figure in regulating lower level phenomena and determining the higher level consequences of lower level phenomena that there will always remain an important empirical task that higher level sciences must perform. They must identify the higher level objects or organizational networks in nature. Using the empirical knowledge of these organizational patterns and whatever knowledge the next lower level science is able to provide as to the components that fit into these structures,[11] the higher level science can hope to account for the phenomena observed at its level and, employing this account, explain what appear as local anomalies at lower levels. Until the organization involved is discovered, however, one cannot hope to explain the phenomena.

One might question my use of the term "discover." If we had a complete physics, why could one not deduce the higher level organizational networks? This may seem a plausible undertaking since higher level entities are constituted by lower level entities and their properties are determined by the properties of the lower level objects and the way they are put together. Clearly, though, just having all the lower level laws would not be sufficient—one needs to know the initial conditions of the system. The challenge should therefore be phrased: if one had all the initial conditions specified at the lower level and a complete set of laws governing the behavior of objects at that level, would that not enable one to deduce all the changes in the higher level organizational networks and their consequences? It is rather easy to show that, as a practical matter, such a task is intractable. We already are aware of the difficulties in solving such problems as the generalized three body problem in physics or the multi-locus, multi-allele problem in genetics. In order to make such problems tractable at all we have to fix certain constraints empirically and solve only the more specialized problem defined by the constraints. Calculating, on the basis of the behavior of the lower components of a system, how the various higher level configurations will change over time would be a problem much more difficult than these already insoluble ones. So, as a practical matter, it seems plausible that we will always have to use empirical research to identify the higher level causal networks operating at any one time.

I want to argue, however, that there is a stronger sense in which, given a complete lower level science and statement of initial conditions, there remains an empirical task. The causal effects higher levels have on lower levels prevents us from simply building up from lower level sciences to higher level ones. In order to see more clearly what I am arguing, let me first present the condition in which no empirical task of discovery would remain. That is the case of the Laplacean demon, which has a complete state description of the *universe* and a complete statement of the laws of physics governing the *universe*. If such a demon had sufficient computing power and all the laws were determinative, it clearly could calculate the behavior of objects at every later moment in time. One way the demon might hope to accomplish this feat is to first deduce laws telling how complex systems within the universe respond to the various possible states of affairs accruing inside and outside of them and then to use these laws to calculate the behavior of the objects under the actual conditions that accrue at any given moment. If the demon pursued this approach, what it would be doing would be deriving all the laws of the higher sciences.

Whether this approach is possible depends upon whether there is a complete set of possible modes of organization that could occur in the universe. However, it is implausible to think that the number of possible forms of organization is so limited. Although I cannot give a definitive argument for this claim, the seeming endless variety of organisms and social structures that have evolved from a limited number of components make it seem plausible that, just as there are an infinite number of possible English sentences, there is an infinite number of possible organized systems. If there are, however, even the Laplacean demon is left with an incompletable task, for the demon must calculate separately the effects each new organized system has on its components whenever it reaches the point where that system appears.

Even if science were incompletable for the Laplacean demon, because it could not derive a complete set of theorems governing all possible organized systems, the demon would not require further empirical discoveries. The task before the demon, however, is awesome in a way that is not generally recognized. What the demon must do is not work out individually all pairwise interactions of objects in the universe; rather it must calculate how each pair of objects interacts in light of the constraints imposed by all organized units in which the two objects occur. This means looking at each pair of objects not in isolation but in terms of their interactions with various other objects. This additional complexity in the demon's task makes it unlikely to be what is intended by an appeal to working out a complete account of nature on the basis of a complete physics. What is typically envisioned is that one could study independently this operation of individual entities and account fully for their responses to sets of inputs. One then assumes that if one could do this for each part, one could build up to a complete account of systems built out of these entities. But this does not follow. The reason is one we have already seen. If an entity is part of an organized system, its response to a particular input is partly determined by the system. If the system is organized as to be self-regulative, it can change the environment of the entity in question and so alter the entity's response capacities. This is not to deny that we can learn why the entity responded differently to later stimuli by learning how the changed environment caused internal changes. Rather, the point is that without knowing the design of the system in which the entity is embedded one will not understand how the environment changes. To understand that one needs to know how the system is organized. So, for organized self-regulating systems, knowing all there is about the internal operation of a part (or of all the parts) is inadequate to explain the behavior of the part (or of the total system). One also needs knowledge of the organization and of the principles governing how, in view of the particular mode of organization, the conditions affecting the behavior of each part will be altered. Thus, one cannot start by taking the parts independently and explain their response to a series of inputs and then develop an explanation of the whole system by accumulation of parts explanations.

This problem is a well known problem with *in vitro* experiments in physiology where one tries to study the response characteristics of isolated components. In these *in vitro* studies one removes all the effects of the organization in the system and so fails to discover how the part actually behaves in *in vivo* conditions. By adding together what is learned from studying all the parts in isolation, one still fails to explain the *in vivo* system. Only by adding in knowledge of the organization of the system does one come to understand the regulative properties of the system and so reach a position where one can explain the *in vivo* behavior of the system.

This latter case differs from the Laplacean demon case in that in the Laplacean demon case it was assumed that all the interactions of the parts in the total universe were assumed to be accounted for in explaining each entity's behavior; whereas in this case only the principles governing the internal operation of the entity are taken into account when explaining how the entity responds to inputs. The difference between the two cases is fundamental. While both cases require abilities to calculate that are beyond human limits, the requirements on a Laplacean demon are so beyond human reach that we do not know how to approximate it in our research. The second programme, on the other hand, is tractable. We can isolate and study the response patterns of individual entities. But, whereas the Laplacean case does hold the promise of deducing the behavior of complex systems from the principles governing the parts, the second alternative does not. The difference is that the Laplacean demon begins with complete knowledge of the overall organization and of how each part is affected by it. From this, the demon calculates how the parts behave and how the organization changes. In the more realistic case, however, by isolating the parts for study, one eliminates from one's study the effects of organization. All regulative effects of the organization on a component are looked at as simply further inputs. The order of these inputs is lost and can only be recovered by determining the organization within which the entity is embedded. This is the sense in which one must empirically discover what organized system an entity is embedded in, if any, and determine its effects on the entity before one can completely explain the entity's behavior. By putting together knowledge of the organization and knowledge of the principles governing the behavior of the parts one can hope to produce a more complete explanation of the system's behavior, although that endeavor may be limited if the system itself is embedded in a still higher level system.

I have presented the case of building from parts explanation to explanation of wholes to show the significance of organization, something frequently overlooked in reductionistic research programmes (Wimsatt, 1980). But, in fact, even if one allows for empirical discovery of modes of organization and their effects, this programme is not practical. *In vitro* experiments never actually give us complete knowledge of the operation of the entity such that we know how it would respond to every possible environmental changes. Rather, we gain only partial knowledge of the entity as we try to figure out how it responds to particular stimuli. Hence, early on in actual research we need knowledge of the actual organization in which it occurs in order to figure out what features of the entity need to be understood. What we need to account for in our models is how the entity responds to the kinds of local environments it actually encounters. Thus, in practical scientific endeavors, knowledge of organization must enter at an early stage of inquiry in order to make the attempt to explain a part's operation tractable. Unfortunately, this is often ignored by empirical scientists who subject a component to test conditions they assume are relevant without attending to the organization in which the entity is embedded. (As I try to show in several forthcoming publications, the history of science provides a catalogue of errors which such research has produced).

CONCLUSION

I have argued that the hierarchical ontology of nature makes the scientific enterprise incompletable. The behavior of objects at each level in this hierarchy will be partly determined by the organizational structure in which they are embedded. It is through being embedded in these organizational networks that lower level objects come to constitute higher level ones. These organizational networks not only determine the properties of the higher level objects, but, through feedback systems, affect the behavior of the constituent objects. These constituents will thus behave differently in the special environment created by the organizational network than when free in nature or when bound in another object. This undermines the attempt to build a complete explanatory science by just learning how individual entities at any one level behave. One must discover how these entities are affected by higher level organization. I have suggested that even a Laplacean demon could never calculate all the possible modes of organization. For us, they must be discovered empirically. Moreover, as a practical matter, we should not postpone this higher level investigation until we complete the lower level one since what we actually need at the lower level is primarily knowledge of how components will behave under the conditions found in the organized structure that constitutes the next level in the hierarchy of nature.[12]

NOTES

1. Although the issue is often put as to whether a complete physics is the only science one needs, I shall speak in a more general fashion, asking whether a complete science at any level is sufficient. I will clarify the notion of level in section 2 below, but for now one can understand a level as comprised of the objects and events studied by a particular science. The hierarchy of levels I discuss, then, corresponds to the hierarchy of sciences (e.g., physics, chemistry, biology, psychology, sociology).

2. As Robert Richardson has pointed out (personal communication), this characterization will not do. Following Davidson's (1970) strategy, even a dualist could use the predicates of physics to pick out mental events (e.g., "the thought I had at the same instant as a particular stellar collision"). The problem here is generated by the attempt to use the analysis of language to settle ontological issues. Below I will reject the attempt to understand reductionism in linguistic terms and suggest using a part-whole type analysis. In that framework, one can readily establish the ontological claims of physicalism—higher level entities are comprised solely of lower level physical objects. I shall also show how something like Fodor's rejection of the explanatory primacy of the lower level follows within my account.

3. Fodor too recognizes the possibility of using lower level phenomena to explain higher level phenomena in contexts where type-type identity cannot be established, but denies that this allows for reduction. One reason for this is perhaps that, for Fodor, reduction is eliminative reduction. It is clear that without type-type identities one does not have eliminative reduction—the only way one could do without higher level laws is to embed all of them in the lower level laws. But that is elimination in name only.

4. For example, the molecules of a billiard ball interact with another billiard ball only in the context of being parts of the first billiard ball. It is in terms of the whole billiard ball that one analyzes the interaction. Moreover, in ontological terms, the molecules constituting the first billiard ball are not like random molecules in the atmosphere. They are integrated into one object and so act together.

5. In a paper forthcoming in *Philosophical Studies*, I indicate how one can use the part-whole relation in understanding the relation between phenomena at different levels. I argue that identity is not the appropriate relationship between objects at different levels and suggest that Pylyshyn's notion of a functional architecture gives us a better tool for understanding how neurological events come to realize cognitive properties.

6. I have discusses this historical case at some length in Bechtel (1982a). I also argue that it is by attending to organizational features of the nervous system that materialists can

hope to explain features of human mental behavior that have led dualists to posit a non-material mind. Campbell (1974) has also discussed the regulatory effect of higher order organization on component parts under the label "top-down causation." In Bechtel (forthcoming *b*) I extend the argument to show that the environment in which a cognitive system is embedded may be crucial for determining the intentional or psychological ascriptions we make to it and so conclude that we should treat intentional attributions as involving real relations between a cognitive system and its environment.

7. This is part of the issue that lies behind the units of selection controversy in biology. Gene selectionists, arguing that all selection forces must affect gene frequencies if they are to affect evolution, hold that the gene level is primary (Dawkins, 1976). One of the most compelling arguments against the gene selectionist view is that being in a group may differentially affect fitness, so that if one wants to understand the actual cause of evolution, not just to project over average tendencies, one must look also at higher level groups as units of selection as well (Wimsatt, 1980). As Hull (1978) has argued, populations have sufficient internal cohesiveness to act like individuals, and so are the right kinds of entities to experience differential survival and reproduction. Wade (1978) has shown that a major reason population model builders have been able to deny the causal efficacy of higher levels of selection is that their mathematical models employ assumptions that deny that the entities at these higher levels have sufficient integrity to have causal efficacy. One of their assumptions is tantamount to assuming blending inheritance at the group level which, as Jenkins points out against Darwin, undermines the inheritance of variation needed for selection to have anything to operate on.

8. This, incidentally, is part of the insight that is needed to answer charges by creationists that the organizational network found in living organisms is too improbable to occur by chance. If stable sub-assemblies can be built, and the whole composed of such sub-assemblies, then the improbability is much reduced. We are not concerned with the improbability of all parts coming together at once. In addition to this insight, one needs to remember that there was no reason that evolution had to end up with what now exists—but some outcome (possibly one without life) was necessary. Each of these outcomes is improbable just as a particular series of coin flips is. But when one flips a coin twenty times, one gets such an improbable sequence. These two points together provide a coherent answer to the creationist's charge about the improbability of evolution, for the near decomposibility argument reduces the improbability of the observed outcome while the second argument shows that any outcome would have been an improbable one.

9. Cognitive psychology risks serious errors insofar as it insists on methodological autonomy both from inquiries into the ecological niche in which a subject exists and into the neurological mechanisms that realize cognitive states. In the paper mentioned in Footnote 5, I try to show how the ontology I have sketched in this paper allows for a legitimate autonomy to psychology (it describes interactions occurring at one level in the hierarchy of nature) but allows for using knowledge from other levels of inquiry to show the constraints on psychological activity.

10. Only gradually have biochemists come to attend to other kinds of organization within living cells. In Bechtel and Ecanow (forthcoming) we argue that physical-chemical organization within cells is a significant factor which holds potential for explaining many unexplained or problematic cell phenomena like differential ion concentrations, anesthesia, and gas transfer. Biochemists, however, systematically ignore such effects of physical-chemical organization and so are forced into positing such entities as energy consuming ion pumps for which insufficient energy is available.

11. As will be clear below, I am not suggesting that one must have a completed account at one level to develop the account at the other level. Historically, the need to explain a particular higher level phenomenon has often given the impetus for and provided insight to the study of lower level processes. Such study has, in turn, forced revisions in higher level theories. More generally, studies at different levels are needed to guide and correct study at any one level. A model for this type of integration of research effects at different levels can be found in Darden and Mauld's (1977) account of interfield theories.

12. My thanks to Adele A. Abrahamsen and Robert C. Richardson for their most helpful comments on an earlier draft of this paper and to members of the audience at the Center for the Philosophy of Science (especially Carl Hempel, Ron Giere, and James Fetzer) for thought-provoking points that they raised.

REFERENCES

Bechtel, W. (1982 *a*). "Taking Vitalism and Dualism Seriously: Toward A More Adequate Materialism," *Nature and System*, vol. 4, pp. 23–43.

Bechtel, W. (1982 *b*), "Two Common Errors in Explaining Biological and Psychological Phenomena," *Philosophy of Science*, vol. 49.

Campbell, D. T. (1974). " 'Downward Causation' in Hierarchically Organised Biological Systems." in F. J. Ayala and T. Dobzhansky (eds.) *Studies in the Philosophy of Biology* (Berkeley: University of California Press).

Darden, L. and Maull, N. (1977). "Interfield Theories," *Philosophy of Science*, vol. 44, pp. 43–64.

Davidson, D. (1970). "Mental Events," in L. Foster and J. W. Swanson (eds.), *Experience and Theory* (Amherst: University of Massachusetts).

Dawkins, R. (1976). *The Selfish Gene* (Oxford: Oxford University Press).

Fodor, J. A. (1974). "Special Sciences," *Synthese*, vol. 28, pp. 97–115.

Fodor, J. A. (1979). *The Language of Thought* (Cambridge: Harvard University Press).

Fruton, J. S. (1972). *Molecules and Life: Historical Essays on the Interplay of Chemistry and Biology* (New York: J. Wiley and Sons).

Hopkins, F. G. (1938). "Biological Thought and Chemical Thought: A Plea for Unification," Linacre Lecture. Reprinted in J. Needham and E. Baldwin (eds.), *Hopkins and Biochemistry*: 1861–1947 (Cambridge: Wheffe and Sons, 1949).

Hull, D. L. (1978). "A Matter of Individuality," *Philosophy of Science*, vol. 45, pp. 335–360.

Nagel, E. (1960). *The Structure of Science* (New York: Harcourt, Brace).

Richardson, R. D. (1979). "Functionalism and Reductionism," *Philosophy of Science*, vol. 46, pp. 533–558.

Richardson, R. C. (1982). "How *Not* to Reduce a Functional Psychology," *Philosophy of Science*, vol. 49, pp. 125–137.

Schaffner, K. F. (1967). "Approaches to Reduction," *Philosophy of Science*, vol. 34, pp. 137–147.

Schaffner, K. F. (1980). "Theory Structure in the Biomedical Sciences," *Journal of Medicine and Philosophy*, vol. 5, pp. 57–97.

Simon, H. A. (1969). *The Sciences of the Artificial* (Cambridge: M.I.T. Press).

Wade, M. J. (1978). "A Critical Review of the Models of Group Selection," *Quarterly Review of Biology*, vol. 53, pp. 101–114.

Williams, M. B. (1970). "Deducing the Consequences of Evolution: A Mathematical Model," *Journal of Theoretical Biology*, vol. 29, pp. 343–385.

Wimsatt, W. C. (1976). "Reductionism, Levels of Organization, and the Mind-Body Problem," in G. Globus, G. Maxwell, and I. Savodnik (eds.), *Consciousness and the Brain: A Scientific and Philosophical Inquiry* (New York: Plenum Press).

Wimsatt, W. C. (1978). "Reduction and Reductionism," in P. D. Asquith and H. Kyberg (eds.), *Current Research in Philosophy of Science* (East Lansing: Philosophy of Science Association).

Wimsatt, W. C. (1980). "Reductionistic Research Strategies and Their Biases in the Units of Selection Controversy," in T. Nickles (ed.), *Scientific Discovery: Case Studies* (Dordrecht: Reidel).

THE LIMITS OF ECONOMIC SCIENCE

Daniel M. Hausman

When beset with fundamental doubts about the value of their achievements, economists look longingly to philosophy for help. No doubt scientists in other disciplines fiddle with philosophy of science too. They oversimplify it in introductory lectures and in introductory chapters of textbooks. They throw it in one another's teeth in the heat of polemical battles. They think seriously about it when they confront new problems that seem peculiar or confusing. But the self-doubts of economists are exceptional. They are only equalled by those that torment the practitioners of other, even less developed social sciences.

When times are hard for economic theory—as they are now—(see Bell and Kristol, 1981), economists start boning up on their philosophy of science, because they hope that they will be able to trace their difficulties to some simple, but definite methodological error or errors. Economists would, ironically, like nothing better than to discover that they have been methodological boneheads! In that case their ailments would have a definite cure.

By praising the methodological practices of economists, one crushes their hopes for such a philosophical "quick-fix." One also shows that scientific methods have their limits. Such praise for economists should not be exaggerated. One can, if one wishes, compile a hair-raising catalogue of stupendous methodological stupidities endorsed by and committed by economists. But economists have no monopoly on such gaffs. Einstein and Bohr could revolutionize physics while expounding (but probably not putting into practice) untenable methodological views. I shall argue that the inadequacies of economics are not mainly due to the methodological blunders of its practitioners, but to the recalcitrance of its subject matter. Scientific method has simply not served economists well.

On first glance this assertion might appear to be a variety of "anti-social scientific naturalism" (see Morgenbesser 1970, p. 20). Hundreds of authors have argued that social phenomena like those studied by economists lie beyond the limits of scientific method—or at least beyond the method of the natural sciences. Such anti-naturalists have maintained that the methods of the natural sciences *cannot* apply to the study of societies and that various aspects of social life demand different methods of investigation. This paper contends, on the contrary, only that scientific methods have not worked very well for economists and that they are unlikely to work well. It does not assert that the inefficacy of scientific methods in economics is absolute or that it follows from anything special about human beings. Furthermore, I shall not argue, and I do not believe that there are any better ways to study economic phenomena than to employ scientific methods. I am instead making the more pessimistic claim that these, the best methods of knowledge acquisition, nevertheless have their limits and that one should not expect much of economics.

It is difficult to articulate and defend the thesis that scientific methods have not worked well and are unlikely to work well in economics, since nobody has established any detailed and explicit account of what "scientific method" is. It is thus hard to judge definitely whether economists have in fact followed scientific methods and have, to a considerable extent, failed *nevertheless*. Instead of making any futile attempt to say what scientific method is, I shall focus on the main lines of methodological *criticism* to which economic theory has been subject and shall point out that: (1) those methodological features of economics which have been criticized often differ little from actual characteristics of various natural sciences; (2) that the few real methodological differences between economics and some of the natural sciences can be explained as reasonable responses by economists to specific difficulties posed by the subject matter of economics; and (3) that there is no reason to believe that the failures of economics can be attributed to methodological blunders of economists.

Since its first formulation, standard microeconomic theory has been subject to numerous methodological criticisms. The most important and pervasive of these runs as follows:

> The fundamental theory of neoclassical economics is too "unrealistic." If taken literally and without qualifications, the basic "laws" of economics are simply false. It is not true that all businessmen always attempt to maximize profits or that individuals are never satiated. If qualified with *ceteris paribus* clauses, such claims are insulated from any possible experimental refutation—of which there is not much danger anyway, since economists pay so little attention to testing. Economists instead devote their energies to mathematical efforts to discover what various formulations of their basic "laws" coupled with diverse auxiliary assumptions imply. Demonstrations of the existence and stability of "equilibrium" under a variety of simplified circumstances and comparisons of the properties of different equilibria make up the bulk of standard economics. If one leafs through almost any economics journal, one will find an assortment of models developed with an almost blithe disregard for questions of application or testing.

This general methodological complaint is repeated again and again not only by critics of mainstream economics, like the American institutionalists (Veblen, 1919), but also by orthodox economists themselves when they turn to methodological musings in presidential addresses or in essays on how to do economics (Leontief, 1971; Blaug, 1980, esp. ch. 7).

The above criticism points out four features of mainstream economics and argues that these are in fact four methological *errors*:

(1) Economists rely on generalizations that, if taken literally and without qualifications, are false.

(2) Economists hedge their generalizations with *ceteris paribus* clauses which render them unfalsifiable.

(3) Economists devote little effort to testing their basic theory and are largely unperturbed by apparent refutations.

(4) Economists devote an undue amount of effort to developing abstract and inapplicable mathematical models.

Each of these characteristics of economics calls for detailed discussion, which I have undertaken elsewhere (Hausman, 1981b, esp. ch. 7) and cannot provide in this short paper. But I think that one can show that (1)-(4), are not *errors* at all.

First, the basic "laws" of economics are, if taken literally and without qualifications, false. But so are the laws of virtually all sciences. There will always be possible disturbances and interferences of which theories take no explicit account. To adopt Galileo's strategy and to abstract from the complexities of real-world circumstances is not, in itself, to commit any methodological error. Other tactics *might*, of course, work better in economics. I have, in fact, argued elsewhere that the abstractions and idealizations of orthodox economics are probably too extreme (Hausman, 1981a; 1981b, ch. 7). The orthodox approach may well not be optimal, but it is not unscientific.

But are the results of employing a Galilean strategy in economics usable and testable? The second criticism asserts that the *ceteris paribus* qualifications implicitly attached to the basic "laws" of economics make them unfalsifiable (Hutchison, 1938, pp. 40–46). If "falsifiability" is supposed to be some sort of conclusive affair that employs only deductive logic, then it is quite true that *ceteris paribus* qualifications render generalizations unfalsifiable. But this sort of falsifiability exists, of course, nowhere except in careless formulations of Popper's views. The possibility of making allowances for "disturbances" or "interferences" does not always render generalizations untestable. Otherwise one would never be able to test any scientific laws (Rosenberg, 1976, p. 137). So the problem must lie with the specific *ceteris paribus* qualifications that economists make. But few critics have realized that they need to point out some *special* vice of *ceteris paribus* clauses in economics (Blaug, 1980, pp. 67–69, is a notable exception), and none have succeeded in doing so (Hausman, 1981b, ch. 7).

The claims of economics are particularly difficult to test, and economists have been particularly loathe to test them. As the third criticism above alleges, testing plays a comparatively smaller role in economics than it does in many of the natural sciences. Herein lies a real methodological *difference*. Is it also a methodological *error*? Is orthodox microeconomics in reality a "dogmatic" "metaphysics" (fueled no doubt by ideological pressures) masquerading as science? Some economists who wear Popperian spectacles have made virtually this charge (see for example Hutchison, 1938; or, less pejoratively Bray, 1977).

Economists do little testing for three reasons. First, even though the basic generalizations of economics are certainly not universal laws, economists already know that there is some truth to them (Robbins, 1935, pp. 15, 77; Mill, 1843, Bk. VI, ch. ix). Very little reflection on one's own behavior shows that there is certainly *something* to the claim that individuals have transitive preferences. With only a little further reflection (and there is some experimental evidence as well), it is equally obvious that people's preferences are not always entirely transitive. No test will ever show that the generalization that people's preferences are transitive is universally valid or entirely worthless. Second, economists are generally unable to do controlled experiments. Since it is generally impossible to escape the disturbances and the mess of real economic circumstances, it is extremely difficult to get informative results (Mill, 1843, Bk. VI, ch. vii). Third, even if one were able to get reliable test results, they might well be of little comfort and assistance because of the extent to which economic circumstances are changing (Hicks, 1979, pp. x–xi). Suppose one develops some model, ventures to apply it and makes a "risky" prediction. The prediction is, let us suppose, borne out, and the economist feels that she is on the scent of something true and useful. But by the next time one attempts to apply the model, the relative importance of the various causal factors involved may have changed dramatically. Those factors which are left out of the theory and regarded as merely interferences may have become predominant.

So what should economists do? Spend more time testing their basic theory? For what purpose? How much effort should one expend to generate some largely ephemeral demarcation between those circumstances to which the theory reliably applies and those to which it does not? Perhaps with more testing economists might find out that their basic theory works even more miserably than is now recognized. The result might motivate more economists to look for alternative theories or to seek only relatively superficial generalizations. But such results will never and (if philosophers of science like Kuhn (1970, p. 77), Lakatos (1970, p. 121) and Laudan (1977, pp. 27–30) are to be believed at all) ought never to lead economists to give up a theory with the heuristic and classificational powers of standard microeconomics.

There is, moreover, no compelling reason why economists should attempt to formulate more superficial and directly testable generalizations. Those who have tried to do so, like the nineteenth-century German historical school or the twentieth-century American institutionalists have had little success. There was nothing unscientific about their efforts; they just did not succeed. Given the constantly changing importance of the different causal factors that affect economic phenomena, there is certainly no conclusive reason to believe that more "realistic" generalizations will be valuable and informative. There is nothing unscientific about pursuing an abstract economic theory which is hard to test.

But, finally, one might still question whether economists do not devote unjustifiably great attention to abstract and completely inapplicable mathematical models. Again, I see no methodological *mistake*. In investigating and elaborating the implications of the basic laws they have discovered, physicists have explored various models, even when they knew that the basic claims of the models could not possibly be true of nature. Mainstream orthodox economists have attempted to follow exactly this strategy.

The strategy has not worked very well. Galilean abstraction thrives on the possibility of doing controlled experiments (or on the good fortune of being able to study a virtually closed system like the solar system). Unable to perform controlled experiments and cursed with a nasty messy subject matter, economists have not gotten very far. They have not been entirely in the dark, though, since casual experience and even introspection give one good reason to believe that the basic "laws" of economics, although problematic, are not entirely worthless. Economists have hoped, through the articulation of ever more complicated models, eventually to achieve a theory with enough scope and structure that its predictions will be clearly (although probably only intermittantly) visible through the incessant interferences and complications of economic phenomena.

So I see nothing unscientific in the overall strategy of mainstream theorizing. At least in regard to its reliance on "unrealistic generalizations," economics is not greatly different from the natural sciences methodologically. Its limited success does not result from any fundamental methodological *error*. There are some real methodological *differences*—especially the relative unimportance of testing in economics. But these differences, as well as the failures of economics result from difficulties inherent in the particular subject matter of economics. We have referred to three such difficulties: (i) economists are generally unable to perform controlled experiments; (ii) the subject matter of economics is "complex"—a large number of different *kinds* of causal factors influence economic phenomena; and (iii) the subject matter of economics is changing—the relative importance of different kinds of causal factors differs at different times.

A much more popular explanation for the limited success of scientific method in economics relies on the fact that economics concerns human behavior. In a sense I agree: the fact economics concerns human behavior contributes to the complexity and changeability of the subject matter and to the difficulties of performing experiments in economics. Moreover, as we shall shortly see, the fact that economists are concerned with the nature and consequences of human actions makes these difficulties particularly hard to avoid. Yet to claim that scientific methods do not succeed in economics because economics is concerned with people is quite misleading. It suggests that there is something peculiar about people— free will, perhaps—which is beyond scientific study. Scientific methods are unlikely to work well on any complex and changeable subject matter on which experiments cannot be performed—regardless of whether human actions are involved.

In pointing to these peculiarities of the subject matter of economics to explain why scientific method has not served economists very well, I am not claiming that those who study the natural sciences never face similar difficulties. Astronomers, geologists, meteorologists and oceanographers are also unable to perform many controlled experiments; yet astronomers and geologists have far soldier achievements to their credit than do economists. Some biologists must confront problems concerning complexity and change which are similar to those faced by economists. Confronted by all these difficulties simultaneously, economists merely find the cards badly stacked against them.

There remains, however, a fundamental philosophical objection to my entire line of reasoning, which needs to be discussed. As philosophers and economists have often stressed, the complexity or the changeability of a subject matter depends on the means of classification that one possesses and thus on the theories one accepts. The gross motions of the planets appeared much more complicated before Kepler and Newton than they did afterwards. In the same way, it is possible that in ten years, having developed a much better theory, theorists will find that economies are affected by few variables and that the relative importance of these influences changes very little. So there is no argument here against the *possibility* that scientific methods will succeed.

But this argument is weaker than it may appear. One is still entitled, given what one currently knows, to make estimates of how likely it is that economics and its subject matter will be thus transformed, and one may still reasonably maintain that the specific difficulties posed by the subject matter of economics (as that subject matter is currently conceptualized and understood) explain why economists have had so little success. Such explanations are not empty. Of course every unsuccessful science can complain about the complexity of its subject matter. But not every unsuccessful scientist can, like the economist, point to such a large number of different *kinds* of relevant causal factors nor to such evident and gross changes in the importance of different causal factors over time as those that economists have observed. These features of economic phenomena may evaporate with the development of better economic theory, but they do not merely indicate theoretical failure.

There are, in fact, two further reasons why it is particularly unlikely that scientific methods will work any better in the future for economists than they have in the past. First, the enormous pressure economists encounter to produce immediately usable results militates against unfamiliar or "revolutionary" conceptualizations of economic phenomena. An esoteric theory that divided up the phenomena which theorists now regard as "economic" in an entirely different way would be unlikely to have any immediate predictive pay-offs. Second, economists face in practice a serious restriction on the sort of explanations they can offer. Since economic phenomena result from or are constituted by ordinary human actions under various constraints, one expects those actions to be described and explained in relatively familiar and recognizeable ways (Machlup, 1955, pp. 16–17). An economist does not, of course, make any methodological *mistake* if he or she depicts human action in some novel way (just as a theorist in cognitive psychology makes no methodo-logical mistake if he or she boldly suggests that there are no such things as beliefs — see, for example, Churchland, 1981). But to break with our ordinary way of con-ceiving of human action is extraordinarily difficult. There are thus strong sanctions which lead economists to continue to talk about familiar and important phenomena in familiar terms.

Provided that economists do not abandon the attempt to explain and to predict things like unemployment or inflation and to do so in terms of human actions within institutional constraints, they cannot escape the familiar ways of classifying eco-nomic phenomena. And until economic phenomena are radically reconceived, their complexity and unsteadiness will remain. That complexity and unsteadiness along with the inability to perform controlled experiments explain why economists have had such limited success. In principle there is no reason why scientific methods might not triumph in economics. In practice, given the nature of the subject matter of economics, they are unlikely to. In confronting a subject matter like economies, scientific method has been impotent.[1]

NOTES

1. I would like to thank Paul Thagard for comments on an earlier draft of this paper.

REFERENCES

Bell, C. and Kristol, I. (eds.) (1981). *The Crisis in Economic Theory* (New York: Basic Books).

Blaug, M. (1980). *The Methodology of Economics* (Cambridge: Cambridge University Press).

Bray, J. (1977). "The Logic of Scientific Method in Economics," *Journal of Economic Studies*, vol. 4, pp. 1–28.

Churchland, P. (1981). "Eliminative Materialism and Propositional Attitudes," *Journal of Philosophy*, vol. 78, pp. 67–90.

Hausman, D. (1981a). "Are General Equilibrium Theories Explanatory?" In *Philosophy in Economics*, ed. by J. Pitt (Dordrecht: Reidel), pp. 17–32.

———. (1981b). *Capital, Profits and Prices: An Essay in the Philosophy of Economics* (New York: Columbia University Press).

Hicks, J. (1979). *Causality in Economics* (New York: Basic Books).

Hutchison, T. (1938). *The Significance and Basic Postulates of Economics*, reprinted (1960) (New York: A. M. Kelley).

Kuhn, T. (1970). *The Structure of Scientific Revolutions*, Second Edition (Chicago: University of Chicago Press).

Lakatos, I. (1970). "Falsification and the Methodology of Scientific Research Programmes," in *Criticism and the Growth of Knowledge*. ed. by I. Lakatos and A. Musgrave (London: Cambridge University Press).

Laudan, L. (1977). *Progress and its Problems* (Berkeley: University of California Press).

Leontief, W. (1971). "Theoretical Assumptions and Nonobserved Facts," *American Economic Review*, vol. 61, pp. 1–7.

Machlup, F. (1955). "The Problem of Verification in Economics," *Southern Economic Journal*, vol. 22, pp. 1-21.

Mill, J. S. (1843). *A System of Logic*, reprinted (1949) (London: Longmans).

Morgenbesser, S. (1970). "Is it a Science?" in *Sociological Theory and Philosophical Analysis* ed. by D. Emmett and A. MacIntyre (New York: Macmillan), pp. 20-35.

Robbins, L. (1935). *An Essay on the Nature and Significance of Economic Science* second edition (London: Macmillan).

Rosenberg, A. (1976). *Microeconomic Laws: A Philosophical Analysis* (Pittsburgh: University of Pittsburgh Press).

Veblen, T. (1919). "Why Economics is not an Evolutionary Science," in *The Place of Science in Modern Civilization and Other Essays*, reprinted in Lerner, M., ed. (1949), *The Portable Veblen* (New York: Viking Press), pp. 215-40.

EXPERIMENTER BIAS AND THE BIASED EXPERIMENTAL PARADIGM

Morris L. Shames

S cience has found sufficient cause for trembling ever since Robert Rosenthal (1958; Fode, 1960; Rosenthal & Fode, 1961) experimentally demonstrated that an experimenter's predilections exert a significant—albeit unintended and presumably, unconscious—influence on the outcome of his research. The dilation of this research into the limits of scientific inquiry in general was founded on this original study; however, it has long since superseded its own experimental constraints. It has been suggested that no one in the scientific community is exempt from what is generally described as "experimenter effects" (Rosenthal, 1966; 1976), yet a number of questions must be satisfactorily answered before this critique of science passes muster. These questions are: (1) How validly can one generalize from research, which is almost exclusively based on the null hypothesis testing paradigm notwithstanding its myriad studies, to science in general with its considerably more diverse methodological armamentary? (2) Does the extant body of experimenter bias research measure up particularly well to the basic canons of experimental research, that is to say, does the general avoidance of the most pristine form of control reflect a distinct form of tendentiousness in the experimental paradigm itself? (3) How much stock can one place in a body of research which impugns the scientific process, in general, and the research process, in particular, when it, paradoxically, employs the same experimental syntax and is, presumably, subject to this self-same impugnment?

THE QUESTION OF GENERALIZABILITY

It has been postulated, based upon an empirically-derived schema, that there are experimenter effects which do not influence an experimental subject's behavior and there are experimenter effects which do—yet in both instances the experimenter impinges on his research so as to systematically affect its outcome (Rosenthal, 1966). We are told that observer effects, interpreter effects and intentional error are sufficiently pervasive as to have permeated the fabric of the physical sciences, the biological sciences and the behavioral sciences. Moreover, the experimenter's biosocial and psychosocial attributes, situational factors such as an experimenter's experience, the effects of the experimenter *qua* model, and the experimenter's expectations in respect of his research are all unintended, yet systematic, determinants of his research.

This is, plainly, cutting a fairly wide swath in the limits to scientific research but before such an asseveration can be accepted, the problem of external validity (Campbell & Stanley, 1963) must needs be addressed. Where the data proffered apropos of experimenter effects are not anecdotal, they are usually derived from experiments performed primarily within the purview of the behavioral sciences (Rosenthal, 1966; 1976). What is common to all of these studies is the experimental paradigm employed, the prototype of which is the classic, person-perception study of experimenter expectancy effects. Here, a principal investigator leads one-half of his experimenters to believe that they will obtain one type of behavior (an average +5 rating in photograph judgment) from the subjects they run while the other half is led to expect the opposite result (an average −5 rating from their subjects). Inasmuch as there are no pre-existing systematic differences in both groups of experimental subjects, the experimenter expectancy effect, or experimenter bias effect, is inferred from the differences in results obtained between both groups of experimenters. This, it is clear, is a null hypothesis testing procedure and is common to most, if not all, of the research comprising experimenter effects.

The question then arises: If claims of generality, in respect of scientific epistemology, are made in behalf of experimenter effects notwithstanding the fact that this research into the limitations of science has been founded on a single, experimental paradigm and the armamentary of scientific method, in general, is far more comprehensive, how probatively useful is this research in defining the limits of scientific knowledge? Insofar as Physics, for example, is concerned with the prediction of quantitative magnitudes and the form of functions rather than establishing its case by means of rejection of the null hypothesis, a procedure, in any case, which is thought to be theoretically unsound and logically fallacious in that it insists upon affirming the consequent (Bakan, 1966; Lykken, 1968; Meehl, 1967; Rozeboom, 1960), and insofar as Physics and Psychology, for example, suffer opposing fates as a result of the same type of improvement in experimental precision (Meehl, 1967), how much verisimilitude can we impute to experimenter effects? Large areas of Biological and Zoological research have long resisted the statistical approach, in general, and the null hypothesis testing procedure, in particular, and this, too, calls into question the minimal external validity of research into experimenter effects and its impugnment of the scientific process.

The multiplicity of diverse paradigms of inquiry in science, in general, is the rule (Kuhn, 1970). This supports the case for both "interscientific," that is to say, interdisciplinary, and "intrascientific" diversity and, therefore, no form of radical holism in epistemology is warranted (Glymour, 1980). Yet notwithstanding this widely accepted position, it is generally assumed that experimenter effects are an affliction plaguing all of science even though the experimental evidence to that effect has been monolithic. This appears to be a case of *a dicto secundum quid ad dictum simpliciter* and, as such, refels the case to be made for experimenter effects as a limiting condition in experimental inquiry.

THE QUESTION OF CONTROL

The point of departure in the question of control is Rosenthal's (1976) own asseveration that:

> The expectations of the scientist are likely to affect the choice of experimental design and procedure in such a way as to increase the likelihood that his expectation or hypothesis will be supported. That is as it should be. No scientist would select intentionally a procedure likely to show his hypothesis in error. If he could too easily think of procedures that would show this, he would be likely to revise his hypothesis (pp. 127-128).

This might indeed be a veridical reading of the extant situation but is this truly "as it should be"? Indeed some have argued that the inverse approach should obtain (Popper, 1959) so that "laws are working assumptions which have been rigorously tested by conscious attempts at falsification" (Peters, 1965, p. 754). In fact, Peters goes so far as to aver that "all that is required for an enquiry to be a theoretical science is that conscious attempts should be made to overthrow hypotheses" (p. 165).

The question then arises: How do Rosenthal and the research he has engendered fare in respect of this problem of experimental design? The answer to this question necessarily comprises an understanding of the concept of control itself, control groups, in particular, and how they relate to the prototypical experimental design in Rosenthal's (1976) research and the research which has issued therefrom.

The use of the control experiment is comparatively early dating from Pascal's attempt to determine the weight of air; however, control, in the sense of a check or comparison in scientific inquiry is a more modern phenomenon dating from late nineteenth-century biological research and early twentieth-century psychological research (Boring, 1954; 1969). Mill (1843) first systematically treated the problem in terms of his canons of induction but the first thoroughgoing account of control groups in psychological research was undertaken by Richard Solomon in 1949. In our time, however, the use of control groups in experimental research has become so commonplace that it becomes a cause for suspicion when it is not fully used in any research undertaking.

As described earlier, the classic experimenter bias design involves the induction of positive and negative biases in groups of experimenters who, in turn, transmit—

as yet, by means not fully known—these expectancies to the naive subjects they run in a person-perception task. There is control in this design, in the broadest sense in that there are comparison groups; however, there is no control group in the pristine sense that a control group is "a group which is equivalent to the experimental group in every respect except for the independent variable which the experimental group is treated with and the control group is not" (Wolman, 1973, p. 163). In order to qualify for the latter a totally unbiased group of experimenters would have to be included in the experimental design.

This is not an idle charge brought against this research in that it reflects an error of commission rather than an error of omission. Moreover, it is suggested that this prototypical design reflects in itself, as Rosenthal suggests but was, himself, unable to guard against, the operation of the bias process and, as such, again, casts doubt on the compass of experimenter effects, especially experimenter expectancy effects, as a broad limitation of the process of scientific inquiry. Thus, if we were to take seriously Popper's criterion of falsifiability, or Peters' notion that a good experiment should, as much as is reasonable, load the deck against itself, this prototypical experiment does not pass muster.

From a more evidentiary perspective, what outcome could one plausibly expect if a more appositely pristine paradigm were to be employed? A study by Burnham (cited in Rosenthal, 1976) affords some insights in that he found that lesioned rats performed a T-maze discrimination problem more poorly than unlesioned (sham surgery) rats, a comparison groups design not unlike the experimenter bias paradigm in form. This, in itself, is not unexpected but when fuller controls were employed, that is to say, expectancy control groups, it was found that the effects of the experimenters' biases, in respect of the animals' performance, were greater than those owing to the lesions themselves. The absence of full controls, then, has the net effect of overestimating the effect of the independent variable—a lesson which one would expect bias researchers to have fully learned.

Research in this area has proliferated enormously in the past two decades and the overwhelming majority of studies has failed to take into account this problem of noninclusion of an unbiased control group against which the induction of bias could be measured in directional terms. This is not surprising since in psychological research previous studies serve as the experimental point of departure for succeeding research which thereby inherits the flaws of its predecessors. Since Rosenthal's research has been both programmatic as well as prototypical in this area, and since more than 85% of the twenty-one studies for which he has had major responsibility have not incorporated such a control group, it is entirely within the realm of reason that the vast corpus of research which has issued from these studies should evidence a similar bias in the fabrication of its experimental paradigm.

A number of additional conclusions may be drawn from the scrutiny of more than 300 studies in this area. First, it should be noted that approximately one-third of these studies has yielded statistically significant (p < .05) results and since

Meehl (1967) points out that significant findings under conditions of the null hypothesis testing procedure do not carry as much weight as disconfirming instances owing to a weaker, if not invalid form of inference involved, then the weight of evidence is massively opposed to the veridicality of experimenter effects. Add to this the problem of control and its impingement on directionality, that is to say, many of the statistically significant findings have included positive ratings in both groups of experimenters, one of which is led to expect a positive (+5) bias in the judgments of their subjects and the other, a negative (−5) average rating, and, again, one is left with the impression that experimenter effects have been sizeably overestimated in the literature. This difficulty with the literature is further compounded when one scrutinizes those studies that assiduously employed non-expectancy control groups (Anderson & Rosenthal, 1968; Barber, Calverley, Forgione, McPeake, Chaves, & Bowen, 1969; Pflugrath, 1962). In each of these instances, it should be noted, no evidence for significant bias effects was obtained.

Taking these data into account, then, along with the suggestion that experimenter effects are both unintended and unconsciously perpetrated, for the most part, it is not entirely surprising that the experimental paradigm has, in most instances, taken the form it has. It is, after all, the ineluctable outcome of formulating a hypothesis for test; however, this does suggest that experimenter effects may be more the resultant of a biased experimental design than experimenter bias.

THE EXPECTANCY PARADOX

The final, and perhaps most grave impugnment of experimenter bias effects arises from the "expectancy paradox" (Shames, 1979). This paradox is founded on the recognition that experimenter bias research is metamethodological in that the research process is employed to investigate the research process itself and this metamethodology, no less than any other method, is constrained by experimenter effects, in general, and the experimenter bias (expectancy) effect, in particular. It follows from this that the more evidence there is for the generality of experimenter bias effects, indeed, the more ineluctable are these effects, then the more pernicious is the expectancy paradox. In the limiting case of absolute inexorability, so too is the expectancy paradox inexorable, and it follows therefrom that investigations into the research process itself become impossibly hampered by the serious factor of indeterminacy. If, on the other hand, experimenter bias effects are found to be considerably less general in nature, then the expectancy paradox poses a far less serious threat to this research but, by the same token, the effects of the experimenter's expectancy are more trivialized.

Two considerations must needs be attended to in dealing with the expectancy paradox: "(1) the inexorability of the experimenter expectancy effect, that is to say, *how much* of any particular research outcome is attributable to this phenomenon, and (2) its generality in psychological research" (Shames, 1979, p. 384). Several reviews of the extant literature bear on this point. In two separate reviews of 311 studies, classified according to eight research areas — reaction time studies, inkblot tests, animal learning, psychophysical judgments, laboratory interviews, learning and ability situations, the prototypical person perception task, and everyday situations — Rosenthal (1976) found proportions of statistically significant studies ranging from .20 (inkblot tests) to .89 (animal learning). The overall proportion of statistically significant studies was in the order of approximately one-third. This, it should be clear, reflects the limited external validity of the experimenter bias effect both within and across research areas of activity.

When this lack of generality is considered, again, in conjunction with Meehl's postulation of the logical assymetry between the formally invalid confirmation of a theory by a prediction employing the null hypothesis testing procedure and its valid form of disconfirmation by the *modus tollens*, the purview of the experimenter bias effect is massively restricted.

SUMMARY AND CONCLUSIONS

Research in the area of experimenter effects purports to impose itself as a source of contamination if not a limiting condition in respect of scientific inquiry, comprising the behavioral, biological and physical sciences. However, the experimental paradigm upon which this asseveration is based has been a person perception task employing the null hypothesis testing procedure. Since scientific inquiry ranges much farther afield than this restricted form of the hypothetico-deductive method, the claims made for experimenter effects are considerably exaggerated. Moreover, the classic research in this area most frequently employs comparison groups, not a methodologically pristine control group as this kind of research must, and it therefore reveals itself as the product of a biased research undertaking. Finally, the research in the area of experimenter bias effects is itself subject to such an effect and, thus, it faces an inescapable paradox. However, since it has been shown that bias effects are neither general nor inexorable, both the expectancy paradox as well as the experimenter expectancy effect are not as troublesome as was suspected.

It would seem that experimenter effects and experimenter bias research do not — as is claimed — sound the tocsin for the limitations on scientific method and, by implication, scientific knowledge in general. This research, by dint of the monolithic paradigm employed despite its voluminous, programmatic character, does not fare well, as has been shown, in its own domain — the behavioral sciences. Certainly the impugnment of its claims applies *a fortiori* to scientific epistemology in general. Where this research has been anecdotal — and much of it has been, especially outside of the behavioral sciences — it does not at all weigh heavily on the limitations of scientific knowledge; where it has not been anecdotal, it should be treated with the sound scientific skepticism it deserves for the very reasons outlined in the corpus of this paper.

REFERENCES

Anderson, D. F. & Rosenthal, R. (1968). "Some Effects of Interpersonal Expectancy and Social Interaction on Institutionalized Retarded Children," *Proceedings of the 76th Annual Convention of the A.P.A.*, pp. 479-480.

Bakan, D. (1966). "The Test of Significance in Psychological Research," *Psychological Bulletin*, vol. 66, pp. 423-437.

Barber, T. X., Calverley, D. S., Forgione, A., McPeake, J. D., Chaves, J. F., & Bowen, B. (1969). "Five Attempts to Replicate the Experimenter Bias Effect," *Journal of Consulting and Clinical Psychology*, vol. 33, pp. 1-6.

Boring, E. G. (1954). "The Nature and History of Experimental Control," *American Journal of Psychology*, vol. 67, pp. 573-589.

Boring, E. G. (1969). "Perspective: Artifact and control," in R. Rosenthal & R. L. Rosnow (eds.), *Artifact in Behavioral Research* (New York: Academic Press, 1969).

Campbell, D. T. & Stanley, J. C. (1963). *Experimental and Quasi-Experimental Designs for Research* (Chicago: Rand McNally College Publishing Company).

Fode, K. L. (1960). *The Effect of Non-visual and Non-verbal Interaction on Experimenter Bias*, unpublished M. A. Thesis, University of North Dakota, Grand Forks.

Glymour, C. (1980), *Theory and Evidence* (Princeton, New Jersey: Princeton University Press).

Kuhn, T. S. (1970). *The Structure of Scientific Revolutions*, 2nd ed. (Chicago: The University of Chicago Press).

Lykken, D. T. (1968). "Statistical Significance in Psychological Research," *Psychological Bulletin*, vol. 70, pp. 151-159.

Meehl, P. (1967). "Theory-testing in Psychology and Physics: A Methodological Paradox," *Philosophy of Science*, vol. 34, pp. 103-115.

Peters, R. S. (ed.) (1965). *Brett's History of Psychology* (Cambridge, Massachusetts: The M.I.T. Press).

Pflugrath, J. (1962). *Examiner Influence in a Group Testing Situation With Particular Reference to Examiner Bias*, unpublished M. A. Thesis, University of North Dakota, Grand Forks.

Popper, K. (1959). *The Logic of Scientific Discovery* (New York: Basic Books).

Rosenthal, R. (1958). "Projection, Excitement and Unconscious Experimenter Bias," *American Psychologist*, vol. 13, pp. 345-346 (Abstract).

Rosenthal, R. (1966). *Experimenter Effects in Behavioral Research* (New York: Appleton-Century-Crofts).

Rosenthal, R. (1976). *Experimenter Effects in Behavioral Research: Enlarged edition* (New York: Irvington Publishers, Inc.).

Rosenthal, R. & Fode, K. L. (1961). "The Problem of Experimenter Outcome-Bias," in *Series Research in Social Psychology*, ed. by D. P. Ray, Symposia studies series, no. 8 (Washington, D.C.: National Institute of Social and Behavioral Science).

Rozeboom, W. W. (1960). "The Fallacy of the Null-Hypothesis Significance Test," *Psychological Bulletin*, vol. 67, pp. 416-428.

Shames, M. L. (1979). "On the Metamethodological Dimension of the 'Expectancy Paradox'," *Philosophy of Science*, vol. 46, pp. 382-388.

Solomon, R. L. (1949). "An Extension of Control Group Design," *Psychological Bulletin*, vol. 46, pp. 137-150.

Wolman, B. B. (1973). *Dictionary of Behavioral Science* (New York: Van Nostrand Reinhold Company).

CONTRIBUTORS

DAVIS BAIRD, Assistant Professor of Philosophy
    University of South Carolina
WILLIAM BECHTEL, Assistant Professor of Philosophy
    University of Illinois / Medical Center
STEPHEN G. BRUSH, Professor of History
    University of Maryland
JOSEPH L. ESPOSITO, Editor, *Nature and System*
JAMES H. FETZER, Visiting Associate Professor of Philosophy
    New College of the University of South Florida
DANIEL M. HAUSMAN, Assistant Professor of Philosophy
    University of Maryland
HENRY E. KYBURG, JR., Burbank Professor of Moral and Intellectual Philosophy
    University of Rochester
EARL R. MacCORMAC, Charles A. Dana Professor
    Davidson College
MORRIS L. SHAMES, Associate Professor of Psychology
    Concordia University
DENNIS TEMPLE, Associate Professor of Philosophy
    Roosevelt University

INDEX OF NAMES